WHO PUT THE BUTTER
IN BUTTERFLY?

WHO PUT THE BUTTER IN BUTTERFLY?

David Feldman

Illustrated by Kassie Schwan

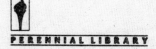

PERENNIAL LIBRARY

Harper & Row, Publishers, New York
Grand Rapids, Philadelphia, St. Louis, San Francisco
London, Singapore, Sydney, Tokyo, Toronto

Who Put the Butter in Butterfly?, read by the author, is available on audiocassette from Harper Audio, a division of Harper & Row, Publishers, Inc.

A hardcover edition of this book was published in 1989 by Harper & Row, Publishers.

First PERENNIAL LIBRARY edition published 1990.

The Library of Congress has catalogued the hardcover edition as follows:
Feldman, David, 1950-
    Who put the butter in butterfly?
    Bibliography: p.
    Includes index.
    1. English language—Etymology. 2. English language
—Terms and phrases.
PE1574.F45 1989   422   88-45575
ISBN 0-06-016072-1

ISBN 0-06-091661-3 (pbk.)
01 00   RRD 20 19

For the original Penguins—Jon Blees, Larry Prussin, Bill Stranger, and Kent Beyer—who taught me the meaning of the words WAH and BITL.

Contents

Acknowledgments

Shortly after I signed my first book contract at Harper & Row, my editor, Rick Kot, took me to meet Barbara Rittenhouse and Mark Landau in the Special Marketing department. Although it was a friendly visit at first, the three of them proceeded to tie me to an easy chair with some spare hemp that was lying around Barbara's office. At first I feared violence, but soon I found out that the purpose of the abduction was to elucidate the Master Plan the three of them had charted for me. You are now reading step one of the Master Plan.

This book would not have been written if Rick, Barbara, and Mark hadn't generously given me the idea. Nor could it have been written if they hadn't, in a compassionate moment, untied me.

Others at Harper & Row have helped me enormously, without the need for physical props. Publisher Bill Shinker has been encouraging and enthusiastic. Everyone in the Special Marketing and Publicity departments has been wonderful to me. Special thanks to Joann diGennaro, who arranged my last publicity tour and lived to tell about it. Elisheva Urbas has helped with so many problems, big and small, with such good cheer, that my hair has temporarily stopped graying. Connie Levinson has been a constant source of support and humor.

Thanks, as always, to my friend, agent, and man of the world, Jim Trupin; to Kas Schwan, for her terrific illustrations and friendship; to Mark Kohut, for his encouragement and counsel; and to my friends and family, for their love.

ACKNOWLEDGMENTS

Preface

When I wrote *Imponderables* ™, I purposely omitted any questions about the origins of words and phrases. An Imponderable was a mystery that couldn't be answered by standard reference books, and so many wonderful books about words already existed that I couldn't believe there was a need for more.

As someone who uses words for a living, I have long wondered why I so effortlessly spew words or clichés whose origins I know nothing about. After being inundated with questions about word and phrase origins from readers of the *Imponderables* series and callers on talk shows, I realized I was not alone. This book is an attempt to sate your curiosity and mine.

Unlike *Imponderables* books, which demand original research, this project required what a teacher of mine called "book work." I'm indebted to the many word-lovers and scholars, from Charles Earle Funk, William Safire, John Ciardi, and William and Mary Morris, to the editors of the *Oxford English Dictionary*, for their groundbreaking and often painstaking research. If you are interested in pursuing the fascinating field of etymology, you will find a treasure trove in the bibliography included here.

One reminder that will make this book easier to understand. You will often run into the phrase "first recorded." Many words start as folk expressions, and it may be decades or even longer before they are committed to print. Scholars use "first recorded" to identify when a given word or phrase was first printed, without denying the possibility that the phrase was used earlier in the spoken language.

If you are itching to find out the story behind the origins of a favorite word or expression that isn't contained here, the last page of this book will explain how to unburden yourself of the affliction and win a free autographed book as well.

But for now, *chill out* and *keep your hat on.*

Clichés and Other Words
to the Wise

Why Is Mincing Around a Subject Called *Beating Around the Bush?*

Medieval man may not have had the thrill of flinging Frisbees, but they had a worthy counterpart, the challenging sport of *batfowling*. A rare nocturnal sport, batfowling consisted of going into a forest or shrub-laden area and beating birds senseless with a bat.

Batfowlers sought sleeping birds for their prey, but being true sportsmen, they didn't want to kill a defenseless bird. So before whacking it with the bat, they were kind enough to wake the bird up first, by stunning it with a harsh light, rendering the bird blind and temporarily helpless. "Sensitive" batfowlers caught the birds in nets rather than using the Darryl Strawberry approach.

Sometimes, though, the birds proved to be uncooperative, selfishly sleeping in bushes where they were invisible, instead of marching forward and offering themselves as ritual sacrifices. So batfowlers engaged servants or boys, known as beaters, to literally beat adjacent bushes to rouse flocks of sleeping birds. As the stunned birds awakened and fled in panic, they would be attracted to the torch or lantern and be socked into unconsciousness by the batfowler.

Although the person today who beats around the bush might not have violence on his mind, he similarly conceals or avoids the real thing that concerns him. While he might pretend to be interested in the bush, he might be more interested in the bird, or worm, lurking inside.

Why Does *All Wool and a Yard Wide* Mean "Genuine"?

Fraud in the marketplace is hardly a twentieth-century invention. In 1464 England passed a law regulating fraudulent practices in the selling of woolen materials. Disreputable fabric salesmen still foisted off adulterated products as "pure wool" and shortchanged customers on measurements in the late nineteenth century, for this is when this phrase was coined. *All wool and a yard wide* originally meant nothing more than that a customer was receiving what was promised.

Why Does *Back and Fill* Mean "to Vacillate"?

Back and fill always sounded more like a disco step than its actual meaning, which has a long nautical tradition. In sailing,

DAVID FELDMAN

backing means to let the wind blow sails against the mast. *Filling* means to let the wind blow the sails toward the bow. *Backing and filling* means alternating having the sails "filled" with wind and then allowing the wind to escape by hauling "back" on the stays.

Yes, backing and filling impedes the movement of the ship, but sometimes this is necessary. When tacking a ship, a navigator might want to keep the boat in the same place. If the tide is running with the ship but the wind is against her, backing and filling is the usual tactic to steady the boat, even if this results in alternating movements forward and backward. Backing and filling is also a way to let the tide take control of the movement of a boat, especially when negotiating through narrow channels or rivers where banks, wharfs, or other objects stand as dangerous obstacles.

Why Do We Say an Outlaw Is *Beyond the Pale?*

In the twelfth century, the Norman conquerors of England decided to set their sights on neighboring Ireland. They managed to capture much of the area around Dublin and some other coastal cities. For protection from Irish attacks, the Normans (later, the English) fenced off their property with *pales* (from the latin *palus*) or "stakes."

The region around Dublin became known as "the pale," and *pale* became a noun signifying any territory. The expression *beyond the pale* was originally applied to an untamed Irishman but was clearly popularized by the Rudyard Kipling story of the same name.

BELL
OUT OF
ORDER

PLEASE
KNOCK

Why Do We Say *Knock on Wood* Instead of *Knock on Aluminum* or *Knock on Naugahyde?*

We live in an age where it is easier to knock on Formica or "genuine" simulated walnut paneling than real wood. But ancient civilizations didn't have the benefit of our technology, and many of them worshiped trees. North American Indians circa 2000 B.C. venerated oak trees. Many pre-Christian Eu-

DAVID FELDMAN

ropean cultures also believed that guardian spirits resided in trees. Knocking on wood (i.e., on a tree), then, was an attempt to wake up the spirits and enlist their help.

Many other claimants have lined up to protest that *knock on wood* was their creation. Among them:

1. The Irish. Knocking on a tree was the signal to thank leprechauns for their help.

2. Christians. The argument goes that the *wood* referred to in the saying is the wood on the cross upon which Jesus was crucified. Knocking on wood, then, is a way of connecting with Christ and recognizing his fate.

3. Jews. During the Spanish Inquisition, synagogues served as refuges as well as places of worship. Jews developed an elaborate knocking system to ensure their safe entrance and egress from synagogues. If one "knocked wood" properly, one was safe.

4. Tag players. In some variations of tag (called both "wood tag" and "tree tag"), if you touch a tree, you are "free." The English say "touch wood" rather than "knock wood," and tag might be the source of the English variant.

The many alternate theories of word origins inevitably lead to ethnocentrism. We assume that an expression must have been created in the same context in which we have first experienced it. Although the four explanations above are quite different, they all put a spin on the ritual of the ancients who believed that good spirits lived inside trees.

Submitted by Herbert Israel of West Palm Beach, Florida.

What Was the First Stuff to Be *Cut and Dried?*

Herbs. Physicians in the sixteenth century were likely to prescribe herbs as treatment for most maladies. Although they also were used as cooking ingredients, much faith was put in herbs as "modern" remedies. Physicians preferred dispensing herbs that were already *cut and dried,* because dry herbs are both more concentrated and more uniform in strength.

While one batch of fresh thyme might vary in potency from another, dried thyme made the dosage routine, which, come to think of it, is pretty much our definition of *cut and dried* today.

Why Do We Say That Someone Who Is Finally Concentrating on Serious Business Is *Getting Down to Brass Tacks?*

In the nineteenth century, most tacks were made of copper, but not those found in English fabric stores. Retailers placed brass tacks on the inner edges of their sales counters, exactly one yard apart. When a customer finished browsing and selected a skein of cloth or other fabric, she was literally *getting down to brass tacks*—ready to measure the length of fabric and pay for it.

The brass tacks later yielded to a brass rule built into the edge of counters. The ruler was obviously more accurate in measuring lengths less than one yard.

What Are *Hackles* and Why Do They Get Raised?

Hackles are feathers on the neck of a rooster or peacock. The expression was first recorded in 1883 by writer Edward Pennell: "I almost saw the hackles of a good old squire rise as he waved his hat and cheered."

The origins of "raise hackles" is easy to speculate on. Much like the hairs on the back of a dog's neck, the hackles of a rooster or peacock rise when they get agitated.

Why Does *the Jig Is Up* Mean "Your Game Is Over"?

This expression has nothing to do with the dance, and its roots are English (first recorded in 1592) rather than Scottish or Irish. *Jig*, in medieval times, was slang for "trick." *The jig's up*, a classic line delivered by hard-boiled detectives to exposed con men, has had the same meaning for four hundred years.

Which Is the *Last Straw?*

It's the same straw that broke the camel's back. Actually, *last straw* is a variation of Archbishop John Bramhall's (1655): "It is the last feather that breaks the horse's back."

DAVID FELDMAN

Charles Dickens probably is responsible for the spread of the cliché *last straw*, as it appears in *Dombey and Son*, his 1848 novel: "As the last straw breaks the laden camel's back, this piece of underground information crushed the sinking spirits of Mr. Dombey."

Why Does *Knuckle Under* Mean "to Submit" or "to Give In"?

Knuckle once referred to the joint of any bone, including the knee and elbow. *Knuckling under* originally meant "to bend down on one's knee and kneel in submission." The knuckles of the knees were hitting the ground and under (and supporting) the body of the supplicant.

Why Does *Knuckle Down* Mean "to Concentrate" or "to Work Hard"?

This time, *knuckle* does refer to the knuckles of the hand. This English expression is about 250 years old and originally described the game of marbles! In marbles, if a player keeps the knuckles of his hand down on the ground, he is concentrating to the fullest extent. The *OED* supplies the first recorded use of the expression: "*Knuckle* or *knuckle down* is a particular phrase used by lads at a play called *taw* [a word that now describes the fancy 'shooter' marble but that was the name of the game in the eighteenth century], wherein they frequently say, 'A *nuckle down to your taw,*' or fit your hand exactly in the place where your marble lies."

Why Is a Final Effort Called *Last Ditch?*

Anyone who fought in a war has probably hoped that the trench he was in was the "last ditch" he would ever see. Most of us would guess that the *ditch* referred to in *last ditch* is a military trench rather than a farmer's irrigation ditch, but few realize that this expression predates the two world wars.

The first recorded use of *last ditch* was in Bishop Gilbert Burnet's memoirs, *History of My Own Time,* published in the early eighteenth century: "There was a sure way never to see it [Holland] lost, and that was to die in the last ditch." The earliest use of *last ditch* was a literal one, signifying a last stand, a last defense against an aggressive enemy.

The first American citation was in a proclamation issued by the citizens of Westmoreland, Virginia, in 1798: ". . . but one additional Obligation, To Die in the Last Ditch or uphold our nation."

Thomas Jefferson was perhaps the first to use the phrase figuratively (1821): "a government . . . driven to the last ditch by the universal call for liberty."

Why Do We Say That Someone in a State of Anxiety or Suspense Is *on Tenterhooks?*

After fabric was woven and milled, the material used to be stretched on a frame, called a *tenter*, to dry the cloth evenly. The cloth was kept snugly in place by hooks (or bent nails), which were known as, appropriately enough, *tenterhooks.*

Tenter probably derives from the Latin *tendere* ("to stretch"). *On tenterhooks*, then, always has been a metaphor for being "on the rack." Novelist Tobias Smollett so used *tenterhooks* as early as 1748: "I left him upon the tenterhooks of impatient uncertainty."

Why Do We Say *Once in a Blue Moon?*

Although myths that the moon was made of green cheese, the moon was blue, and the saying *once in a blue moon* all appeared in the sixteenth century, there is no evidence that anybody took these ideas seriously. A sixteenth-century rhyme by William Ray and J. Barlow reeks with irony:

> Yf they saye the mone is belewe
> We must believe that it is true.

Most likely, the original meaning of *once in a blue moon* was "never."

Yet, even twentieth-century folk insist that the moon sometimes *does* appear blue. Some insist that on full moons, the perimeter of the moon is blue; some claim that on crystal-clear nights, or on exceptionally foggy nights, or in areas full of volcanic ash, the moon appears blue. With all of these "blue-moon sightings," the phrase has gradually shifted in time from meaning "never" to "very rarely."

Although "blue moon" is now used to describe the second full moon in a calendar year, there is no evidence to indicate that this meaning existed when the phrase was coined.

DAVID FELDMAN

Why Does *Pipe Down* Mean "Shut Up!"?

The original *pipe* in *pipe down* was a boatswain's whistle, and *pipe down* was a signal to sailors that they were dismissed for the day and could go belowdecks. By the late nineteenth century, the tattoo signal, usually sounded immediately before "taps," explicitly meant to quiet down. Obviously the deck was quieter once the sailors went belowdecks. Eventually the boatswain's whistle was replaced by a bugle, but the meaning of the signal was the same.

Why Are the Elite Called the *Upper Crust?*

In the Middle Ages, bread was dispensed, even at formal meals, by diners tearing off chunks from a big loaf. Heaven forbid that the bread of an aristocrat might be touched by the hands of a commoner! So it became the custom to slice off the upper crust of the loaf and present it to royalty (or whoever was the most distinguished person at the table), both as a way of honoring the elite and of keeping out the potential germs of the hoi polloi.

What the Heck Does *Pop Goes the Weasel* Mean?

After I was asked this question by a faithful reader of *Imponderables*, I asked a random sample of my illustrious and well-bred friends this question. None had a clue to the answer. Nor, come to think of it, did I.

Pop the Weasel, it turns out, is not the innocent children's rhyme it appears to be. Its meaning remains elusive to us because it contains some obscure Cockney dialect.

So here, for your edification, is the annotated guide to *Pop Goes the Weasel:*

> Up and down the City Road [1]
> In and out the Eagle [2]
> That's the way the money goes
> Pop [3] goes the Weasel! [4]

Thus our children's rhyme is the inspiring tale of a tailor who blows all of his money on booze and has to hock his equipment to eke out a living.

[1] The City Road is a major thoroughfare in London.

[2] The Eagle was a real pub and popular watering hole in London.

[3] Slang for "pawn."

[4] Slang for a tailor's iron.

Has Anybody Ever Been Given *Long Shrift?*

How many times have you ever seen the word *shrift* without *short* preceding it? Never? *Shrift* is one of those words that I call an "inevitable." As soon as you see it, you know its partner *short* will be alongside it, just as surely as *ample* will be followed by *parking* (was *ample* coined simply as a way of describing large parking lots?).

Shrift has a long history of its own; *scrifan* was Anglo-Saxon for "to receive confession." *Shrift* is simply the noun form of *shrive*, which today means "confession to a priest." We use *shrive*, in particular, to describe the process of giving confession and receiving absolution upon one's deathbed, an important ritual of the Catholic faith.

Short shrift, then, refers to the inability to give a full confession or receive absolution, and the expression comes from the practice of not giving condemned prisoners enough time to shrive properly. Often they would be allowed only a few seconds to speak at the gallows before the executioner. In *Richard III*, Shakespeare alludes to the practice. The Duke of Gloucester, soon to become Richard III, has just sentenced Lord Hastings to death. The official in charge of the execution, Sir Richard Ratcliff, tells the wailing Hastings:

> Dispatch my lord; the duke would be at dinner:
> Make a short shrift; he longs to see your head.

So although we have seen many long shrifts in movies and soap operas, no such expression has evolved, and *short shrift* has broadened its meaning to any context in which we allow little time or give insufficient attention to the matter at hand.

Animals and Other Inhuman Words

Why Is the Middle of Summer Called the *Dog Days*?

No, *dog days* is not an invention of the greeting card industry
to create a phony holiday for your canine pets. Nor is it an

ironic reference to the fact that midsummer isn't exactly Bowser's favorite time of year.

Dog days goes back to the Romans, who believed that in the hottest part of the summer, Sirius (the "dog star" and the brightest star in its constellation) lent its own heat to the heat of the sun (*sirius* means "scorching" in Greek). The Roman *dog days*, which they called *caniculares dies* ("days of the dog"), lasted from approximately July 3 to August 11, when Sirius is ascending. Over time, *dog days* has come to mean any expected long streak of heat.

DAVID FELDMAN

How Did the *Butterfly* Get Its Name?

Despite the simplicity of its etymology (yes, the insect got its name from combining *butter* and *fly*), word experts have been arguing about the genesis of the word *butterfly* for centuries. Samuel Johnson claimed that the season when butterflies first appeared (spring) was when butter was also first churned. Others contended that the *butter* referred to the yellow color of its excrement.

Two other explanations are far more likely. In England, the most common butterfly is the brimstone, which is butter-colored. William and Mary Morris offer a more tantalizing explanation: Medieval folklore tales included the myth that witches and fairies would *fly* and steal *butter* at night—in the form of *butterflies*.

Submitted by Joan Wolf of West Babylon, New York.

What Exactly Are *Cooties,* and Where Did They Come From?

Kids throw around the word *cooties* without having the slightest idea what these vile imaginary creatures are. But since World War I, *cooties* has referred to head and body lice. *Cooties* comes from the Polynesian word for "parasite," *kutu.*

Why Is a Spider's Handicraft Called a *Cobweb?*

Cob is a short word with disparate meanings. *Cob* can refer to any small lump (e.g., a piece of coal), a horse, a gull, a swan, and, of course, a corncob. But *cob* has nothing whatsoever to do with *cobwebs*.

In Old English, spiders were called *attercoppes*, literally "poison head." Evidently the Anglo-Saxons believed that all spiders were poisonous. The word *copweb* appeared in Middle English to describe the net created by spiders, and over the years the spelling changed from *copweb* to *cobweb*. Try pronouncing *copweb* aloud and you will see the effort required to enunciate it properly. The English penchant for slurring words probably explains why *cobweb* has endured.

Submitted by Jean Hanamoto of Morgan Hill, California.

DAVID FELDMAN

Why Can Some People *Get Your Goat* Instead of *Getting Your Mynah Bird* or *Getting Your Basset Hound?*

Although there is some dispute about how this colorful term for the uncanny ability of some people to rile us, annoy us, irritate us, vex us, and get under our skin, most lexicographers attribute the origins of *get your goat* to the world of thoroughbred horse racing. Horse trainers have long put a companion in stalls with high-strung thoroughbreds, particularly volatile stallions.

Putting a horse of the same sex in the stall would lead to territorial battles; putting a horse of the opposite sex in the stall might, to put it politely, distract the stallion from the task at hand. Goats, among the most boring and least demanding of animals, soothed horses effectively.

Horses tended to become attached to their goat roommates, so much so that rival barns sometimes would steal the goat of a rival the night before a race. The horse would become upset and presumably underperform the next day. So someone whose goat has been gotten is actually being compared to a horse rather than a goat.

Why Is a Pedestrian Violation Called *Jaywalking?*

When the colonialists first came to America, blue jays abounded along the Eastern seaboard. As more and more immigrants settled, the jays retreated to the countryside, until eventually *jay* became synonymous with "hick" in the mid-eighteenth century.

Rural dwellers were often dumbfounded by the chaos of big-city traffic. They crossed in the middle of the street, crossed intersections on red lights after traffic signals were invented, and darted out on the street without looking for

DAVID FELDMAN

cross traffic. *Jaywalking* meant "hick-walking." Today we need an antonym for *jaywalking*, a word to express the ruthlessly efficient kamikaze tactics of pedestrians exhibited by big-city urban dwellers.

Submitted by Cynthia King of Morgan Hill, California. Thanks also to Sharon M. Burke of Los Altos, California.

Why Are Some Beetles Called *Ladybugs*?

While recently watching an episode of *High Rollers* (purely for educational purposes, of course), Wink Martindale posed this question:

"True or false: All ladybugs are female."

"False!" said the contestant.

We should hope so, or else these delightful creatures would be incapable of reproducing.

How appropriate that the ladybug should be the one insect with uniformly positive associations, for the "lady" in *ladybug* was Our Lady, the Virgin Mary. Ladybugs were so honored, presumably, because of their contribution in eating harmful insects such as aphids and scale insects that are destructive to plants. Less publicized, however, is the fact that a few types of ladybugs actually eat plants themselves.

In Great Britain, *ladybugs* are called *ladybirds*. In the United States, we might then have to rename them *first ladies*.

Why Is an Outdoor Bazaar Called a *Flea Market*?

The first flea market was held in Paris, but the concept spread throughout Western Europe. Originally, to be a *flea market*, the sale had to be outdoors and the goods secondhand. The assumption was that the old merchandise would gather fleas as well as customers.

Submitted by Lenore Punk of Point Pleasant, New Jersey.

Why Are Politicians Who Have Not Been Reelected but Are Serving Out Their Terms of Office Called *Lame Ducks?*

Politicians are used to worse four-letter words than *lame duck*, but the meaning of these words is bleaker than any obscenity: You are being booted out of office and have to find a *real* job now.

The original *lame ducks* were not ducks and not politicians but financial persons, members of the London Stock Exchange in the eighteenth century, who were clobbered by bears rather than embraced by bulls. Those who could not pay off their debts were booted out of their seats and referred to as *lame ducks*.

By 1863 the expression was used to describe American officeholders, particularly holdover congressmen. The reason a duck was chosen, rather than an aardvark or a penguin, is probably because of a similar old hunter's expression, "Never waste powder on a dead duck." A congressman voted out of office was clearly wounded but far from dead, for although he might have lost an election, he didn't yield his seat until

DAVID FELDMAN

March 4, plenty of time to sail through pork-barrel measures, punish old enemies, and generally create havoc without having to answer to his constituencies.

The problem was so obvious that the Twentieth Amendment was created to resolve it. New congressmen now join the fray in January, and, of course, the Senate and House of Representatives have been totally efficient deliberative bodies ever since.

Why Do We Say *Let the Cat Out of the Bag* Rather Than *Let the Gerbil Out of the Bag?*

Gerbil won't work because this expression is not metaphorical. In medieval times, pigs were sold live at fairs and open markets. Pigs aren't exactly docile, and they don't cotton to standing still in stalls while shoppers eye their potential as bacon.

Without room to house the pigs in pens, the only practical solution was for the sellers to tie up the pigs in burlap sacks. The customers couldn't see what they were buying, and a gullible buyer—the type that today would see a three-card monte game in Times Square as an opportunity to enhance his financial security—might end up buying a cat rather than a pig. When the buyer finally opened his bag, the truth was revealed.

The theory that some buyers were actually fooled is bolstered by another cliché, *a pig in a poke*. A *poke* is Middle English for "sack" or "bag." This expression implies a blind guess, one that is as likely to turn "catty" as "porcine."

Why Does It Rain *Cats and Dogs?* Why Not *Ostriches and Yams?*

We cast our lot, on circumstantial evidence, for the consensus view that this phrase goes back to Norse legends, which contended that animals had specific magical powers. Cats were reputed to have the ability to conjure up storms (visual representations of storms show witches taking the form of cats), and dogs were symbolic of wind. To Scandinavians, then, *raining cats and dogs* meant a violent storm with wind and rain, pretty much what it means to us today.

Holt argues that the phrase probably stems from seventeenth-century England, when Jonathan Swift, in *Polite Conversation,* described the city's gutters as full of debris—including cats and dogs.

DAVID FELDMAN

Why Is the Last Performance or Work of an Artist Called a *Swan Song?*

Just about every author they have *Cliff Notes* for in Classics classes seems to have written about *swan songs.* Plato, Aristotle, Chaucer, Coleridge, Spenser, Shakespeare, and other, less stellar writers have referred to the legend of the dying swan. Although actual swans never sing, they were once believed to sing a beautiful melody just before they died. Socrates attributed the song to a display of happiness at its impending reunion with the god it served. Other ancient myths included that swans accompanied the dead to their final resting place (sort of a reverse stork) and that the souls of dead humans reside in swans.

Because an artistic *swan song* always constitutes the last work of an artist, the allusion to the dying swan is apt. Of course, some artists are not worthy of a first song, let alone a swan song, as Samuel Coleridge commented:

> Swans sing before they die—'twere no bad thing
> Should certain persons die before they sing.

DAVID FELDMAN

Show Biz and All That Jazz

Why Were *Do-Re-Mi-Fa-So-La-Ti-Do* Chosen to Represent the Notes of the Musical Scale?

If it were not for some modifications made in the seventeenth century, the hit song from *The Sound of Music* would have been "Ut-Re-Mi," for our current octave was a modification of a hexachord scale called the "solfeggio system." Invented by Italian Guido D'Arezzo in the eleventh century, the mnemonic used to remember the scale was borrowed from the first syllables of each line of an existing hymn to St. John:

> *Ut queant laxis*
> *Resonare fibris*
> *Mi gestorum*
> *Famuli polluti*
> *Solve polluti*
> *Labii reatum*

When the octave replaced the six-note scale, a seventh note was needed, so the last line of the hymn was appropriated:

Do was later substituted for *ut* because it is more euphonius, and *ti* was substituted for *si* because it is easier to sing. (Try it!)

Is the French *Apache* Dance Named After the American Indians of the Same Name?

Apache came into the French language before the dance; it was used to describe Parisian street thugs. The apache dance, then, mimics the violence of criminals rather than the ritual dances of Native Americans.

But how did *apache* ever come to mean "criminal" in French? As if they don't have enough to answer for with their devotion to Jerry Lewis and unlistenable popular music, the French have always been fascinated by James Fenimore Cooper's tales of the Old West. Émile Darsy, a French journalist, was so impressed by Cooper's hyperbolic descriptions of the Apaches' great warriors that he thought that the comparison between them and the local street terrorists was apt. Although the Apaches were fine warriors, they hardly lived up to the malevolent image portrayed in Cooper's work.

Apache Indians got their name from the Zuni word meaning "enemy." Since Apaches obviously didn't see themselves as enemies, they originally called themselves *dene,* meaning "human being."

DAVID FELDMAN

Who or What is the *Allemande* Referred to in the Square Dance Calls *Allemande Left* and *Allemande Right?*

Allemande means "German woman" in French. A French dance called the allemande was popular in the late seventeenth century, and one of its steps quickly insinuated itself into square dancing in French Louisiana.

Why Do We Say *Break a Leg* to an Actor on Opening Night?

Right before an actor goes onstage on opening night, say, "Your makeup is dripping all over your shirt!" or, "By the way, Steven Spielberg is in the front row and he has come to see you," and you are likely to be met with a measure of equanimity. But wish an actor "good luck" and you will be facing one frightened actor.

A "good luck" is perceived by the superstitious acting community as a brazen act of tempting fate, so *break a leg* has come to be the ironic way of wishing "good luck" while stating the opposite. Our expression is a translation of a German expression used for the same purpose, *Hals-und-beinbruch* ("May you break your neck and your leg").

Some have speculated that *break a leg* is a reference to the "unlucky" actor John Wilkes Booth, who managed to break his leg while jumping onstage after assassinating President Abraham Lincoln. But *break a leg* is recorded only in the twentieth century, and, most likely, if Booth were the inspiration, the phrase would have circulated earlier.

Submitted by Joanna Parker of Miami, Florida, and Launie Rountry of Brockton, Massachusetts.

Why Is a Prominent Person or Someone Under Scrutiny Said to Be *in the Limelight?*

Before Thomas Drummond, theatrical productions had one major lighting problem: how to provide strong illumination that featured only one actor or one area of the stage. The solution was a *limelight*, invented in 1826 not to help the theater but installed in a lighthouse near Kent, England, to help guide ships at night.

Lime was a crucial ingredient in utilizing Drummond's lamp. One stream of oxygen and one of hydrogen were burned upon a cylinder of lime. When placed in front of a reflector or behind a lens, *limelight* yielded an intense stream of white light. The limelight provided by far the best way to focus on one actor at a time, and it remained so until the invention of the spotlight.

Euphonious Words—And They Sound Good, Too!

Why Are Laudatory Quotations on a Book Cover Called *Blurbs?*

As a man of letters, I decry and condemn the proliferation of puffery perpetrated by purported professionals. Scan the back cover of just about any hardcover novel and you will see the laudatory quotes from other famous writers praising the "revelation" of the contents within and the "immortal genius" of its creator. A lot of self-congratulatory backslapping if you ask me, and precisely because I've never been able to get even an obscure writer to praise me (even off the record), I am a totally objective observer on this issue.

Blurb, which contains only one letter too many to be a four-letter word, was coined by American humorist and illustrator Gelett Burgess (1866–1951). Burgess is probably best remembered for one of his intentionally humorous pieces of doggerel:

WHO PUT THE BUTTER IN BUTTERFLY?

I never saw a purple cow.
I never hope to see one.
But I can tell you anyhow,
I'd rather see than be one.

Understandably cynical about an industry in which this quatrain could "make" his career, Burgess insistently, but with good humor, lampooned the publishing industry. (He once jokingly defined *blurb* as "to make a noise like a publisher.")

At a dinner party given by the Retail Booksellers' Association in 1907, copies of his new book *Are You a Bromide?* were distributed to the storeowners for the first time. The cover, with a drawing by Burgess, depicted a sickly-sweet young woman and facetious praises of her supposed charms. Burgess named his model Miss Belinda Blurb.

Even in 1907, "blurbs" were prevalent on book covers, even if they hadn't been named yet. Burgess wrote his own parody of the blurbs for his own back cover of *Are You a Bromide?*, and the name of the insipid cover girl was transferred to the intentionally inflated enconiums found on the back of the book.

No doubt Burgess would be sad to find that, if anything, blurbs have proliferated since his death. But he is resting easier knowing that he is now perhaps better known for the word he coined than for his rhyme about the purple cow.

Why Is a Pattern of Crossing Lines Called a *Crisscross*?

Originally, *crisscross* was actually *Christ-Cross*. In the sixteenth century, European education was primitive. Only the three R's and religion were taught to youngsters. Children were taught from a hornbook primer that contained a few numbers for arithmetic, the Lord's Prayer, a few words for spelling lessons, and the complete alphabet. Above the alphabet was a small cross, which was called the *Christ-Cross*.

Right below the *Christ-Cross* was the alphabet line; the alphabet line, because of its proximity to the cross, became known as the *Christ-Cross Row*. Eventually, *Christ-Cross Row* was changed to *Crisscross row*, presumably because the alphabet itself contained no religious content. The change-over is easier to understand when one realizes that the "i" in *Christ-Cross* was a short "i," pronounced like the "i" in *Christmas* and *crisscross* rather than the long "i" in *Christ* or *bicycle*.

Why Is Something Great, a "Real Knockout," Called a *Doozy?*

There is a quaint feel to the word *doozy*, perhaps because the object it was first created to describe has long vanished. The first *doozy* was the Duesenberg, an American car created by two brothers named Duesenberg. State of the art at the time it was produced (1921–37), the Duesenberg was considered more elite than the Cadillac. The engineer brother, Frederick, aimed so high that he installed the same high-performance engines in boats and airplanes as well as automobiles.

What is a *Kit Cat?* And Why Does Every City Seem to Have a Sleazy Dive Called the *Kit Cat Club?*

The first Kit Cat Club was named after Christopher (Kit) Cat, a cook who was a member of London's Whig Club, formed in 1703. Cat originally held meetings in his mutton-pie shop. Although small at first, the Whig Club grew in numbers and stature, as several prominent Londoners joined.

DAVID FELDMAN

Sir Godfrey Kneller painted portraits of the then forty-two members of what came to be known as the Kit Cat Club. He hung the pictures in the club dining room, which was too short to contain the traditional half-size portrait, so Kneller revised the usual format. All forty-two portraits, painted on twenty-eight- by thirty-six-inch formats, were less than half-size, but always included the subjects' hands. Even today, *kit-cat* is the generic name given to this format of portraits.

The Kit Cat Club folded about twenty years after its inception, but it lives on in the name of the American chocolate bar, the glory of having all forty-two portraits hung in London's National Portrait Gallery, and in Kit-Cat bars and strip joints all over the world. Why a once estimable political club lent its name to dives all over the world is unclear, but the alluring alliteration of *Kit Cat* and the fact that the first Kit Cat Club's founder was in the food and drink industry probably had much to do with it.

Why Is a Leader or Boss Called a *Honcho*?

This expression was brought to the United States from fliers stationed in Japan during World War II. In Japanese, *hancho* means "leader of the squad." Most of the time, one hears *honcho* in the expression *head honcho* ("head leader"), which is an obvious redundancy unless the speaker is referring to a boss of bosses (such as a chairman of the board presiding before the top executives of a company).

Why Are Jitters Called the *Heebie Jeebies?*

Heebie jeebies is a coined expression first written in 1910 by a man with an unusual distinction. Cartoonist Billy de Beck, creator of Barney Google, coined two other slang expressions that, though they now sound quaint, have survived for more than half a century: *hotsy totsy* and *horse feathers.* Few novelists or essayists have coined more than one enduring expression; de Beck, with the assistance of Mr. Google, originated three.

Why Does *Hobnob* Mean "to Mingle" or "to Chat Socially"?

Hobnob goes back to the Middle English *habben* ("to have") and *ne habben* ("to have not"). *Hobnob* is a contraction of these two words.

In the twelfth century, when Chaucer used the word, *hobnob* meant "hit or miss" or "give and take" as well as "have and have not."

Hobnob eventually described the age-old custom of alternating purchasing rounds of drinks (literally "having" and then "not having" to buy the next drink). Although the use of *hobnob* is no longer confined to drinking, the conviviality and sociability conjured by *hobnob* resemble the interaction among a group of drinkers.

Does Anyone Ever Engage in *Low Jinks?* And What's a *Jink?*

Originally a Scottish word, the primary meaning of *jink* has, for the past two hundred years, been "to move swiftly, especially with sudden turns." *High jinks* has existed just as long. Although we now use *high jinks* to refer to any prank or frolic, during the late seventeenth century and the eighteenth century, *high jinks* was a popular parlor game. A throw of dice would determine one "victim." He would have to embarrass himself by performing some prank for the amusement of the other revelers; if the victim refused or couldn't satisfactorily perform the task, he had to pay a forfeit, usually downing a hefty container of liquor.

If this is the origin of HIGH *jinks*, it is probably just as well that there is no LOW *jinks* in our lexicon.

Why Is Something Clean and Trim Referred to as *Spic and Span?*

Spic and span is a contraction of a Middle English phrase used to describe a new or refurbished ship. *Spick and span new* referred to a ship with shiny nails (*spics* or "spikes") and new wood (*span new* mean "new chips of wood"). The phrase was shortened eventually to *spick and span* and was forever altered by the commercial success of the cleaner *Spic and Span*.

Why Is Somebody Who Abstains from Alcohol Called a *Teetotaler?*

Teetotal cropped up in England and the United States at about the same time, and it's impossible to discern who first coined it. Dick Turner, an Englishman, felt so strongly that *teetotal* was "his" word that his epitaph commemorates it:

> Beneath this stone are deposited the remains of Richard Turner, author of the word *teetotal* as applied to abstinence from all intoxicating liquors, who departed this life on the twenty-seventh day of October 1846, aged fifty-six years.

DAVID FELDMAN

The early temperance movement in the United States didn't ask members to abstain from all liquor. Hard-drinking Americans were asked to forsake all but beer and wine. But as the movement progressed, it got more militant. *Teetotal*, whether first coined in England or the United States, was almost certainly a play on words, emphasizing the *t* in *total*. Somebody who signed up as a *teetotaler* gave up all alcohol, as opposed to the wishy-washy *O.P.* ("Old Pledge") members, who promised to abstain only from hard liquor.

Let's Get Physical

Why Are Fraudulent Healers Known as *Quacks?*

Poor ducks. It isn't enough that Groucho Marx defamed them in *Horsefeathers* by combating a flirtatious, mewling woman while rowing across a lake with, "Was that you or the duck?" No. Ducks are now blamed for the ministrations of bogus healers.

Quacks is a shortening of the Dutch *kwakzalver*. *Zalf* is Dutch for "salve," so the *kwak* is clearly a reference to early medical pitchmen who sold cure-alls ("salves") by barking, or "kwaking," like a duck.

Why Is Something Set Aside for a Specific Purpose Said to Be *Earmarked?*

This American expression almost certainly stems from the practice of identifying cattle by the markings on their ears. But there is a possibility that *earmarked* is a biblical allusion. In the days of the Old Testament, servants were given an option after six years' service. If the servant chose to stay with his master, the master bore a hole in his ear with an awl, and the servant remained indentured for the rest of his life.

Why Is a Bathroom Sometimes Called a *Head?*

In old sailing ships, lavatories were put in the bow—or head —of the vessels. If one were lucky, waves hitting the bow would serve as the primary means of cleaning the facilities.

But the very earliest ships had no bathrooms at all for crewmen (officers tended to have primitive facilities at the stern). The lowly crew members had to go through contortions to relieve themselves. They went to the head of the ship, clambered over the bulwarks, and urinated or defecated over the edge. Some vessels had holes cut out near the bulwarks and the bowsprit. A few even included seats along the bow. But no indoor plumbing was provided for lowly crewmen— they were forced to hide among the headsails and the riggings to gain some privacy.

Submitted by Ira Goldwyn of Great Neck, New York.

DAVID FELDMAN

Why Do We Say Someone Eager with Anticipation Is *Licking His Chops?*

Although we now think of this expression as slang, it has estimable roots. In Anglo-Saxon, *chops* meant "mouth" or "jaws." Someone who *licked his chops*, then, was someone who, like most animals and some humans we know, licks his mouth and drools in anticipation of a tasty morsel or two.

Old Nellie

Why Is an Old Person Said to Be *Long in the Tooth?*

The first recorded use of *long in the tooth* was in J. C. Snaith's *Love Lane* (1919): "One of the youngest R. A.s [rear admirals] on record, but a bit long in the tooth for the Army."

The meaning is the same today, but the words don't seem to apply to humans. The answer is that *long in the tooth* originally referred only to horses. As horses age, their gums recede. Their teeth don't actually get longer, but they *look* longer. The older the horse, the longer its teeth look.

Why Is Somebody Who Speaks Frankly or Without Reservations Said to *Make No Bones About It?*

The *OED* cites a 1459 reference: "And found that time no bones in the matere." All of the earliest allusions to "no bones" refer to soups and stews. If one encountered a stew with no bones, one could eat freely, without hesitation or trepidation. John Ciardi notes that although this phrase originally meant "to encounter no difficulties," its connotation has changed to mean "to be undeterred by, especially by moral scruples or conscientious reservations."

How Did the *Pap Test* Get Its Name?

The *Pap test* was named after a Greek-born American anatomist and pathologist with an unfortunately long name, George Papanicolaou. His smear test, developed in the 1920s, didn't gain wide acceptance until the 1940s, and it has since saved thousands of women's lives by diagnosing uterine cancer in its curable stages.

How Did *X Rays* Get Their Name?

In 1895, German inventor Wilhelm Roentgen was conducting experiments with the conduction of electrical charges through gases in a vacuum tube. Much to his astonishment, Roentgen observed that radiation passed through objects that were usually opaque. The applications were obvious, but Roentgen didn't understand how or why radiation worked. For this reason, Roentgen named his invention *X Strahlen* ("X ray"). He used "X" as in algebraic formulas, a modest admission that he couldn't explain his own discovery.

Abbreviations and Other Tiny Words

Who Was the First Guy to Be Called *Guy?*

Guy Fawkes, in the early seventeenth century.

You remember what they say about converts? Fawkes converted to Roman Catholicism, and he despised the anti-Catholic reforms instituted by James I of England. In 1605, Fawkes led a conspiracy, called the Gunpowder Plot, to blow up James and the Parliament on its annual ceremonial opening.

The day before the bombing was to take place, Fawkes was arrested right where his gunpowder was stashed. Responding with appropriate mercy, the British tortured Fawkes, which yielded a signed confession implicating his coconspirators. To reward him for this information, Fawkes was allowed to be hanged with his cohorts.

This pleasant story isn't over yet. To mend wounds, Guy Fawkes' execution day was declared a national holiday. A charming custom during this holiday was for children to march through the streets carrying human figures dressed in decrepit clothing. These figures were called *guys* (in "honor" of Guy Fawkes). Because these "guys" were dressed so poorly and haphazardly, *guy* became slang for "a person of odd appearance or dress."

Possibly the first person to use *guy* to mean "fellow" was Mark Twain, in 1872. Even today, *guy* connotes a man without great distinction. Although a *guy* is an O.K. fellow ("a regular guy"), he is unlikely to be listed in *Who's Who*.

Why Is an Army Enlisted Man Called a *G.I.?*

Because Army men and women wear government issue uniforms. *G.I.* sprung up soon after World War I and became a catchphrase during World War II. Originally, *G.I.* was not an affectionate term, but rather a reference to the impersonality of the Army and an expression of contempt for government property.

What's the Difference Between *Hip* and *Hep?* Why Is It Unhip to Say *"Hep"* When It Used to Be Hipper to Say *"Hep"?*

About all we know for sure is that *hep* predated *hip* by at least twenty-five years. By 1903, *hep* meant "informed," or "in the know," and the expression *get hep* circulated as early as 1906. Musicians, particularly black musicians, modified the term (*hep to the jive* was recorded as early as 1925).

DAVID FELDMAN

Where did *hep* come from? One story posits that *hep* honors a Chicago bartender named Joe Hep, who presumably knew his way around the Windy City. The Morrises offer the fascinating theory that *hep* derives from drillmasters who barraged their men with the insistent command "*Hep,* two, three, four," etc. If this supposition is correct, to be *hep* originally meant, literally, "to be in step."

By 1931, *hep* was occasionally pronounced as *hip*. Possibly a regional dialect rendered *hep* unintelligible. Perhaps *hip* was the shortened version of *on the hip,* jazz slang for smoking opium on one's side. We do know that by 1945, *hip* had replaced *hep* among those who used to be hep. After World War II, only a washed-up or spacey bohemian, such as Maynard G. Krebs on *The Dobie Gillis Show,* would be caught dead saying "hep."

The most likely explanation for why *hep* disappeared is that the artists and nonconformists who first coined the term abandoned it as soon as the culture at large embraced it. Likewise, few things angered the counterculture in the 1960s as much as their "own" words (e.g., *groovy, far out*) being co-opted by the mass media. Soon those words were verboten (although both have subsequently reappeared). Once *Time* magazine spots a trend, the hip (or the hep) want to move on to the next big thing.

Why Were World War II Army Rations, Contained in Packets, Called *K* Rations?

Providing all of one's daily nutritional requirements but none of one's aesthetic requirements, *K* rations were probably so named in honor of their inventor, U.S. physiologist Ancel Keys. *K* also was the Army supply code designation for the rations, and some sources indicated that the code letter, rather than the name of Keys, was responsible for *K* rations.

Delicacy prevents me from providing vivid descriptions of the contents of *K* rations. But highlights include an unrecognizable and usually inedible meat/protein source; the

DAVID FELDMAN

world's hardest chocolate (designed not to melt in the sun); and, in every packet, cigarettes! Even to famished soldiers, K rations tasted like *!?#!%*!, but they kept G.I.s alive when refrigeration and cooking facilities were nowhere to be found.

In *A Browser's Dictionary*, John Ciardi best describes the love-hate relationship between the American G.I. and the K ration: "Packed in a wax-coated cardboard box, it passed as a meal, the double function being to sustain the G.I. while making him angry enough to kill."

Why Is the "I" in the Word *I* Capitalized?

After all, the "m" in *me* is not capitalized, and isn't it impolite to capitalize I when the "y" in *you* is stuck in lower case?

Ego turns out not to be the original reason for the capitalized *I*. In Middle English, the first person singular was expressed with *ich*, eventually shortened to *i* in lower case. But printers encountered difficulties setting the lower case *i*. The letter would be dropped unintentionally or run together with the words that followed or preceded it (see "Why Do We Mind Our *p's and q's* Instead of Our *v's and w's?*"). So the original purpose of capitalizing the *I* was to make it stand out from other single letters and provide it with the status as a whole word.

Now the rock star Prince has extended this principle by capitalizing the word *you*, although his spelling, *U*, is a little eccentric. Despite constantly being accused of narcissism, Prince should be commended for his altruistic and egalitarian philosophy of elevating the second person singular to the prestige of first-person-singular capitalization.

Submitted by Sheila Reiss of St. Petersburg, Florida.

Why Is *Pound* Abbreviated as *Lb.?*

In the zodiac, the symbol of Libra is the scales. How appropriate, then, that *lb.* is an abbreviation of the Latin *libra* ("scales") *pondo* ("a pound by weight"). The original *pondo* was a premeasured weight to be placed in one of the two pans, thus providing a standard to apply against other substances of unknown weight.

> *Submitted by Leslie P. Madison of Anaheim, California. Thanks also to Bryan J. Cooper of Ontario, Oregon, and Daniel A. Papcke of Lakewood, Ohio.*

Why Do the English Call a Bathroom a *Loo?*

Before the invention of sewers and indoor plumbing, human waste products were pitched out of windows. Although it has long been considered the height of chivalry for men to walk on the outside while escorting a woman down the sidewalk, I've always wondered whether the custom didn't start with gentlemen trying to distance themselves from flying filth being hurled out of windows.

The French, at least, had the decency to warn innocent pedestrians of their impending peril. Before pitching the filth, they yelled *Gardez l'eau!* ("Beware of the water!"). The Scottish transformed the French expression into "Gardy loo." This wouldn't be the first time that the British have mangled the pronunciation of a French word (which is pronounced like the word *low*), but some etymologists explain the discrepancy by crediting the phrase not to a corruption of *Gardez l'eau* but to an Anglicization of another French phrase, *lieux d'aissance*

DAVID FELDMAN

("room of comfort"), which is closer to the French pronunciation.

One of humankind's basic instincts seems to be the dire need to create euphemisms to describe places where urination and defecation take place. *Lieux d'aissance* is the parent of our *rest room* and *comfort station*, and the spiritual ancestor of *throne, washroom,* and *bathroom.*

Submitted by Ira Goldwyn of Great Neck, New York. Thanks also to John H. Thompson of Glendale, New York.

If *Mrs.* Is an Abbreviation of *Missus,* Why Is There an "r" in It?

Because *Mrs.* isn't an abbreviation of *Missus* but of *Mistress. Mistress* originally referred to a married woman, not a participant in an extramarital affair.

Submitted by Gil Gross of New York, New York.

Why Does *XXX* Mean "Liquor" in Cartoons?

British breweries in the nineteenth century marked their products with Xs to signify the potency of the liquor. One X indicated the weakest brew; and three (and rarely four) Xs signified the strongest. Cartoonists simply borrowed from the British "rating" system.

DAVID FELDMAN

Why Is Something Done Secretly *on the Q.T.?*

This nineteenth-century expression comes from the first and last letters of *quiet*.

Why Is the Symbol for a Prescription ℞?

The ℞ is the sign of the Roman god Jupiter (the patron of medicines). ℞ was an abbreviation of *recipe* (from the Latin *recipere*—"to receive").

The reason that ℞ was atop all prescriptions was that *recipe* meant "take" in Latin, so that "take" preceded all directions to the patient. Even the English word *recipe* originally referred to medical prescriptions, although the connection between formulas for medical purposes and formulas for cooking were then less farfetched, since both used many of the same herbs and spices.

Submitted by Douglas Watkins, Jr., of Hayward, California.

Take Me Away, Please!

Why Were Military Cars (and Now the Line of Chrysler Cars) Called *Jeeps*?

During World War II, the jeep was developed by the United States as its basic military car. Its official designation was *General-Purpose Vehicle*. *Jeep* comes from combining the initial sounds of *General Purpose*. According to *Brewer's Dictionary of Phrase and Fable*, prototypes of the jeep were called *beeps*, *peeps*, and, heaven help us, *blitz buggies*.

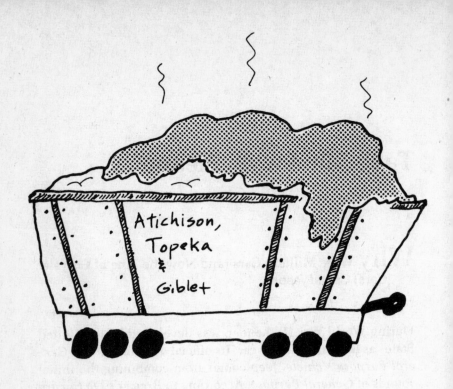

Atichison,
Topeka
&
Giblet

Why Is "Good Money for Little Work" Called the *Gravy Train?*

A literal representation of a *gravy train* conjures up an unappetizing picture; gravy would be most unmanageable cargo, yet the expression has pleasant connotations. *Gravy train* was first recorded in the 1940s in the United States, but as early as the 1930s, *gravy* was slang for "easily obtained money."

Gravy train was borrowed directly from railroad terminology. A *gravy train* was an easy run with decent pay. Like extra, unearned money, gravy isn't necessary to survive, but it sure makes life more pleasurable.

DAVID FELDMAN

Who Is the *Hansom* of *Hansom Cab* Fame?

Come to Think of It, Why Is a *Cab* Called a *Cab*?

Cab is short for the French *cabriolet* ("little leap," so called because the earliest cabs were reputed to be faster than conventional carriages). Cabriolets, traditionally, were one-horse carriages with two seats and folding tops.

Hansom cabs were created by Joseph Hansom (1803–82), a London architect who was bankrupt at the time he designed the cab.

The hansom cab was the immediate predecessor of the classic black London taxicab. By the late 1850s, the hansom cab became a sensation in New York and Boston. Its attraction as a romantic vehicle for scenic areas was guaranteed. The driver, seated behind and above the passengers, afforded privacy to the customers while allowing them an unobstructed view in front. Today, hansom cabs are largely associated with New York's Central Park but still can be found in many other parts of the world.

Why Is a Vehicle That Cuts a Wide Swath or an Object That Attracts Unswerving Devotion Called a *Juggernaut?*

Juggernaut is a Sanskrit word meaning "Lord of the World." In Hindu mythology, *Juggernaut* was the name of an idol to the god Vishnu. Built in the twelfth century, the idol was housed in a temple and brought out every year for the religious "car festival."

Every June or July, the statue is dragged along the street in a thirty-five-foot-square and forty-five-foot-high vehicle. The destination: another temple.

The journey takes several days and does not go unnoticed. Thousands of pilgrims accompany the idol. Some fanatics, in a frenzy, have thrown themselves under one of the car's sixteen wheels.

Why Do We Drive on *Parkways* and Park on *Driveways*? Why Do We Send *Cargo* in a Ship and Send *Shipments* in a Car?

Ever since a popular comedian posed the immortal question "Why do we drive on parkways and park on driveways?" we have been inundated with similar questions. Answering any of these types of queries is a little like pulling off the wings of a butterfly. Any attempt at a logical answer defuses the wit and humor of the question. Let me prove it to you with *cargo*.

All of the words with a *car* root that refer to transportation go back to the Latin *carricare* ("to load on a wagon"), and more particularly to the Roman *carros*, a vehicle used as a baggage wagon by Julius Caesar in his military campaigns. The Spanish coined the word *cargo* to refer to "a burden or load." We'd be the last ever to defuse a joke, but the original cargo was loaded on a cart, the closest Roman equivalent to today's car; in fact, by definition, cargo still can be "shipped" in a car or a truck—*cargo* is merely a synonym for "freight." *Shipments*, although probably originally referring to freight carried by ships, can just as well apply to cargo sent by ship.

And why do we drive on *parkways* and park on *driveways*? One of the main definitions of *way* is "a route or course that is or may be used to go from one place to another." New

York's Robert Moses dubbed his "route or course that was used to go from one place to another" *parkway* because it was lined with trees and lawns in an attempt to simulate the beauty of a park. The *driveway*, just as much as a *highway*, a *freeway*, or a *parkway*, is a path for automobiles. The driveway is a path, a *way*, between the street and a house or garage.

Now can we please have a moratorium on cheap-word-play questions? We warned you this section wasn't going to be fun.

> *Submitted by Peter Vaernet of San Francisco, California, and Sharon M. Burke of Los Altos, California. Thanks also to Brian Hart of Bala Cynwyd, Pennsylvania, and to Ira Goldwyn of Great Neck, New York.*

Why Is the Left-Hand Side of a Ship Facing Forward Called the *Port*?

Why Is the Right-Hand Side of a Ship Facing Forward Called the *Starboard*?

Why Are the Windows Around the Perimeter of a Ship Called *Portholes* When They Are Found on Both Sides of the Vessel?

Ancient ships were usually steered with a large sweep oar. *Starboard* is Old English for "steer board or paddle," so it is easy to see how the right-hand side became known as the *starboard*. Because the steering gear was all contained on the right, old ships had to tie at dock on the left side, and loaded cargo onto the more convenient left, or *port* side as well.

Originally, however, the left-hand side was called the *larboard* (from the Anglo-Saxon *laere* ("empty") and *bord* ("board" or "paddle"). The Middle English *laddeborde* also, appropriately, meant "lading side." In the early seventeenth century, mariners abandoned *larboard* for *port*, undoubtedly because when maneuvering in a wicked storm, *larboard* and *starboard* were too easily confused.

The original purpose for the porthole was not to amuse cruise passengers but to serve as gun ports. In the earliest ships, gun ports were on the port side only. Sailors had primitive, claustrophobic facilities below decks. Portholes as windows were added centuries later, and by that time the word *porthole* had stuck.

Submitted by Robert J. Abrams of Boston, Massachusetts. Thanks also to William DeBuvitz of Bernardsville, New Jersey; John Schroder of Red Lion, Pennsylvania; John Underhill of South Bend, Indiana; and Rich DeWitt of Mound, Minnesota.

When Bad Things Happen to Good Words

Why Is a Free-for-all Called a *Battle Royal?*

Used as early as 1672, *battle royal* was created not to describe a feud among monarchs but rather a slightly less lofty cockfight. Most cockfights were designed as elimination tournaments, starting with sixteen cocks in the first round, eight in the second, four in the semifinals, and the surviving two in the finals. Nobody knows for sure how *royal* became part of the expression, but most likely the sole surviving cock was being compared to royalty—the winner was at the top of the heap.

Battle royal lives today in professional wrestling, where it describes bouts in which ten to twenty grapplers attempt to throw their opponents over the top rope. The last wrestler left inside the squared circle is the momentary king of wrestling.

Why Is Someone Behaving Wildly Said to Be Running *Amok?*

In the sixteenth century, Portuguese traders saw some Malaysians running haphazardly down the streets and attacking people for no apparent reason. Little did the visitors realize that the cause of the frenzy was a nasty strain of opium. The Portuguese took the Malay word *amog* ("engaging furiously in battle"), originally used to describe the ferocity of Malaysian tribesmen, to describe the crazy behavior of these civilians. When in battles, the Malaysian warriors would rush into the front line in a frenzy, engaging in fierce hand-to-hand combat. The tribesmen's ferocity was fueled not only by hatred but by hashish as well.

Why Is Someone Who Is Running *Amok* Said to Be *Berserk?*

Malays hardly had a monopoly on drug-crazed warriors. Although we now stereotype Scandinavians as stoic, ancient Norse warriors terrorized their enemies by charging forward with a suicidal lack of discrimination. What sent them into this frenzy? A prebattle ritual that included eating hallucinogenic mushrooms provided the inspiration.

Berserk means "bear shirt" in Old Norse, and because they battled in bearskins, ancient Norse warriors were called *berserkers.*

In Norse mythology there was an actual warrior named Berserk. Berserk was invincible, and he chose to train his twelve sons to fight as ferociously as he did. Berserkers were commonly referred to as unbeatable and wildly unconven-

DAVID FELDMAN

tional foes. So they were described in *Gods and Myths of Northern Europe:*

> ... frantic as dogs or wolves; they bit their shields and were as strong as bears or boars, they slew men, but neither fire nor iron could hurt them.

Pity the fool who encountered a berserker run amok.

Why Are Disorders or Noisy Demonstrations Called *Bedlam?*

The original *bedlam* was named after a quiet, orderly place, the Priory of St. Mary of Bethlehem, founded in London in 1242. Working-class Londoners, notorious slurrers, pronounced *Bethlehem* "bedlam."

In 1402 the priory was transformed into England's first hospital for the insane. In 1547 Henry VIII established St. Mary's as a royal foundation "for lunatics." When Londoners referred to the Holy City, they took pains to articulate each syllable, specifically to differentiate it from *Bedlam* at a time when mental institutions were equated with chaos and cacophony.

Why Is Nonsense Called *Bunkum?*

If you want your faith restored in American democracy, I urge you *not* to tune in to C-Span's gavel-to-gavel coverage of the House of Representatives. A few hours of interminable, self-serving, pretentious, and incoherent speeches are likely to make you despair for the future of the United States.

Don't worry. Put it in perspective. Windbag speeches are a long-held and cherished tradition in the republic. One windbag, however, redeemed himself and added a word to our language for good measure. In 1820, Felix Walker, a representative from North Carolina, delivered a horrible, rambling, irrelevant, and witless speech. Walker was quite aware that his oration was a disaster. Scores of congressmen called for him to yield, but he refused to do so.

Walker stopped his speech and responded to his fellow Congressmen: "You're not hurting my feelings, gentlemen. I am not speaking for your ears. I am only talking for Buncombe."

"Talking for Buncombe" soon became a catch phrase within Congress and spread quickly outside of the Capitol. *Bunk* and *bunkum,* then, are testaments to that great representative of Buncombe County, North Carolina, Felix Walker.

Why Is Someone with a Hidden Agenda or Selfish Motives Said to Have an *Ax to Grind?*

In "Too Much for Your Whistle," Benjamin Franklin relates a story from his boyhood. A stranger approached young Ben, who sat beside a grindstone. The stranger pretended that he didn't know how to sharpen his ax and asked Ben to demonstrate the grindstone. By the time the man "got it," the ax was sharpened and Franklin was exhausted. If the naive Franklin had been more wary, he would have realized his tormentor had a (metaphorical) *ax to grind.*

Why Do You Say That Someone Who Isn't Worthy *Can't Hold a Candle* to a Worthy Person?

This cliché always struck me as strange. If you can't hold a candle to someone, aren't you doing him a favor? Holding a candle to him could be—well, rather painful. And who would want to hold a candle to someone, and why is someone inferior if he can't?

Note that this phrase always is expressed in the negative. We don't say, "Don Mattingly CAN hold a candle to Wade Boggs." Perhaps this is because at one time there were people who literally "held candles to," and they were far from equal to the people whom they held candles to.

In the sixteenth century, before the advent of streetlights, a common undertaking for servants was to help wealthy Britishers traverse darkened streets. These servants, called *linkboys,* followed their masters on foot while they walked down the street, holding a candle, or more often a *link* ("a torch") to light the way. Linkboys also held candles for their masters in theaters and other public places.

Obviously, the job requirements for linkboys did not include much education or intellectual capacity, so if someone couldn't hold a candle, they were considered to be below the depth of a servant. When we use this cliché in the twentieth century, though, we are left with the original conundrum. Somebody who can't *hold a candle* (i.e., isn't forced to act in a servile role) is deemed, in our language, inferior to the person who can.

To What End Does the *Bitter End* Refer?

The end of a rope. On ships, ropes that are cast a-sea, such as an anchor rope, must be tied to posts, which assure that the rope stays aboard. These posts were called *bitts*. The *bitter* referred to the last portion of the inboard rope attached to the bow bit, and the *bitter's end* referred to the unenviable state of having all the cable paid out with no more room left to maneuver in an emergency.

What Does a *Chowderhead* Have to Do with Soup?

Not much, evidently. *Chowder* is derived from the French *chaudière* ("kettle or cauldron") and the custom of fishermen of throwing different fish into a pot or kettle to provide supper. Stuart Berg Flexner traces *chowderhead*, first recorded in 1848, directly to the soup: ". . . one whose brains were as mixed up as chowder . . . confusion."

But isn't *chowderhead* as likely to be the ancestor of the earlier recorded (1819) English word *cholterheaded* (meaning, naturally enough, "having a jolted head"—i.e., a scrambled brain)?

Why Is Nonsense or Self-Glorifying Blather Called *Claptrap?*

Claquer means "to clap" in French. In the early nineteenth century, evidently, there weren't enough patrons "claquering" away to please French theater owners. In 1820, the estimable M. Sauton established a business to rent out *claquers*, who were hired to applaud a play or, in some cases, a specific actor. The versatile claquer not only had calloused hands but also was adept at laughing uproariously at farces and sobbing copiously at melodramas. Theoretically, the emotional reaction of the audience would appease ego-mad thespians and generate good word of mouth.

The Anglicization originally referred to any trick (i.e., *trap*) to milk applause *(clap)* out of a recalcitrant audience.

Why Are Informers or Strike-Breakers Called *Finks?*

Two equally plausible theories have been advanced to explain the origin of *fink*. The first is that *finks* is a corruption of *Pinks*, short for "Pinkerton men." In 1892, in a bitter dispute with the Carnegie Steel Company, hundreds of demonstrators at the Homestead Steel strike beat and stoned Pinkerton men hired as strike-breakers. *Pinks* became synonymous with "scabs," and *Pinks* was transformed by some genius into the more mellifluous *finks*.

The second, perhaps more likely, explanation is that *fink* was an eponym for one Albert Fink, who over a period of decades became the supervisor of detectives for several different railroads. Fink supervised a team of operatives who were paid to go "undercover" and spy and inform on railway workers. These operatives were dubbed *finks*.

Why Does *Flak* Mean "Strong Opposition"?

Flak sounds like a modern, made-up word, rather than one with an ancient etymological background. And it is. *Flak* was born during World War II and it is formed by the contraction of the first letters of three German words: *FLieger* ("aviator"), *Abwehr* ("defense"), and *Kanonen* ("guns"). So while we now use *flak* to describe any unwanted opposition, the earliest references are to the antiaircraft fire our pilots encountered.

What Pan Was *Flash in the Pan* Named After?

The pan of a flintlock musket. In the seventeenth century, the pan of a musket was where one put the powder that was ignited by the sparks from the flint. If it ignited properly, the sparks would set off the charge in the gun, and this charge would propel the ball (and later, the bullet) out of the barrel.

Occasionally, the priming powder in the pan would burn without igniting the main charge, and the gun misfired. The burn was visible but to no effect, just as a *flash in the pan* is successful but shines only for a brief time.

No musket would discharge unless the *powder was* [kept] *dry*, sage advice that spawned another cliché.

DAVID FELDMAN

Why Is Someone Who Is Depressed Said to Be in a Funk?

In de fonk zun sounds like the name of the new Earth, Wind and Fire album, but it actually is the Flemish expression ("be in the smoke") that has given us our term to mean "in a state of sadness or fear." It was first recorded in an eighteenth-century rhyme:

> Pryce, usually brimful of valour when drunk,
> Now experienced what schoolboys denominate funk.

Why Do We Say That Someone Who Is Fired *Gets the Sack?*

The ancient Romans didn't believe in mollycoddling convicted felons. Rehabilitation wasn't their style. Those convicted of parricide or other heinous murders were tied in a sack and dumped into the Tiber River, instantly solving any potential recidivism problem.

The practice spread throughout many other European countries, and, as late as the nineteenth century, murderers in Turkey were tossed into the Bosporus in a sack. *To get the sack,* then, probably was used figuratively as a threat of any sort of punishment, such as losing one's job.

Another theory to explain how *get the sack* was recorded —as early as 1611 in France—is that it referred to craftsmen of the Middle Ages. Artisans carried their tools in sacks; while they worked, they handed the sacks to their employers. When a craftsman *got the sack,* it meant that his services no longer were required. He was left, literally, *holding the bag.*

Why Does *Corny* Mean "Schmaltzy"?

So many slang expressions, from *jaywalking* to *hick* to *hill-billy*, were weapons for urban folks to put down rural dwellers that it is no surprise that *corny* is a derogatory epithet to describe Americans between the two coasts.

Actually, *corny* is a late-nineteenth-century theatrical expression. When theatrical troops traveled in the hinterlands, they felt smugly superior to their audiences. These thespians, often from New York, felt that "hicks" preferred the most obvious forms of entertainment, especially low comedy and trite and sentimental melodrama. The preferences of these corn-fed audiences soon became known as *corny*.

DAVID FELDMAN

Why Does a Horrible Drug Like *Heroin* Have a "Heroic" Name?

Yes, *heroin* derives from the same Greek word, *heros*, that gave us the English *hero* and *heroine*. Although heroin's manufacture and distribution have long been outlawed in the United States, the morphine derivative was developed as a legitimate painkiller.

Heroin was originally a legitimate trademark taken by a German pharmaceutical company, so the brand name was consciously designed to evoke only positive associations. Not only was heroin effective as a painkiller, it also had the "bonus" of giving patients a euphoric feeling, and as we now know, delusions of grandeur (indeed, it has made many a junkie feel like a hero). Although these side effects can be deadly in an illicit drug, it was at first a distinct selling point in marketing heroin to physicians as a painkiller.

Why Are Whites Sometimes Referred to as *Honkies*, Particularly by Blacks?

After World War II, many southern blacks moved to Chicago, where factory jobs were plentiful. Many of them, living in segregated towns, had little exposure to whites. Many of these blacks' new coworkers were Middle Europeans. In all likelihood, *honky* is a corruption of what was an existing epithet, *hunkie* (or *hunky*), another pejorative term, meaning a lower-class Hungarian. To these blacks, the coworkers they were exposed to were representative of all white people.

Why Do People Say *"I Could Care Less"* When They Obviously Mean *"I Could Not Care Less"*?

Some slang expressions originate in print, but a far greater number start in oral culture and only later are immortalized in print. Such is the story of *I could care less*, which became a catch phrase in the United States after World War II.

Purists rail against this phrase. Isn't it slovenly and insensitive to say or write *I could care less* when one could make one's sentiments absolutely clear by adding the "n't" to "could"?

The usage panel of the *Harper Dictionary of Contemporary Usage* says "thumbs down" to *I could care less*. Only 7 percent of the panel accepts the use of it in writing. Typical of the panel's sentiments is Isaac Asimov's: "I don't know people stupid enough to say this."

In defense of the hoi polloi, *I could care less* is not an illiteracy like *irregardless*. To the 78 percent of the *Harper* panel who refused to accept this expression even in casual speech, we would argue that *I could care less* falls well within the fine tradition of irony. In speech, the emphasis on *could* makes it clear that *could* means "could not."

Think of all the expressions that denote one thing but can mean the opposite with the proper inflection: *big deal, sure, great,* and *right. I could care less* might not mean what it seems, but the meaning, at least in speech, is clear, and the irony enriches our language.

DAVID FELDMAN

Why Is Something or Someone Weak, Insipid, or Sentimental Called *Namby-Pamby?*

Norman Mailer versus Truman Capote. Lillian Hellman versus Mary McCarthy. There's nothing like a good literary feud. And we didn't invent them in the twentieth century.

More than two hundred years ago, prominent poets Joseph Addison and Alexander Pope disagreed on politics and had an intense professional rivalry; acquaintances and friends were forced to choose sides. One of Addison's partisans was Ambrose Philips, a creator of sentimental verse who was somehow convinced that he was a better poet than Pope. The enmity between Philips and Pope was palpable; Samuel Johnson called the relationship between them "a perpetual reciprocation of malevolence."

One of Pope's political allies, Henry Carey, a poet of some repute, wrote a parody of Philips' insipid poetry and called it *Namby Pamby.* Carey took the diminutive of Philips' first name ("Ambrose" became "Amby") and created a rhyme using Philips' last-name initial ("Pamby"). This rhyming technique, called reduplication, is and was common in word games and has spawned many other slang expressions (e.g., *hocus pocus, higgledy-piggledy*). Pope himself later immortalized Philips' (and Carey's expression) by using *namby-pamby* in his own satiric epic *The Dunciad.*

Why Is a Stupid Person Called a *Blockhead*?

A blockhead not only is stupid, but slow and unexciting as well. An idiot, for example, might be an aggressive person who feels he is a genius. For long-standing etymological reasons, a blockhead is likely to be a schlemiel.

Blockhead was first recorded in 1549, and it referred to wooden heads (or blocks) that were used by milliners and

DAVID FELDMAN

wigmakers to display or store their wares. Usually made of yew or oak, most blockheads were in the shape of the human skull, but some were rectangular.

The most famous contemporary blockhead, of course, is Charlie Brown, the protagonist of "Peanuts," and because his head is disproportionately large, some assume that *blockhead* refers to the size of his head. But *fathead* serves that function well. Lucy isn't criticizing him for the girth of his head (after all, studies have shown that powerful people tend to have extremely large heads), but maintaining that poor Charlie has the I.Q. and charisma of a wooden block.

Why Are Sissies Called *Pantywaists?*

Until the 1920s in America, they weren't, but then a type of children's underwear was introduced in which underpants were buttoned to the undershirt. Pantywaists were intended as unisex underwear, but they proved much more popular for girls than for boys. The inevitable result in a sexist society: Boys who wore pantywaists were mocked as effeminate by their peers.

Why Is Rambling, Meaningless Language Called *Rigmarole?*

In 1296, Edward I of England invaded Scotland. As soon as he captured a town, he would force the local elite to sign a parchment document pledging allegiance to their new king.

The document itself was called a *ragman roll,* a then generic name for any official document on a roll, including deeds, wills, and tax lists.

Recitations of the roll were about as fascinating as readings of phone books, so when the Britisher's slovenly speech transformed *ragman roll* into *rigmarole,* they were as stupefied by obscurantist language as we are today.

How Did *Son of a Gun* Become a Euphemism for *Son of a Bitch?*

Actually, *son of a gun* isn't a euphemism for *son of a bitch* at all. *Son of a gun* dates back to the early nineteenth century and was hardly the good-natured endearment it is today. Originally, it was a pejorative reference to a sailor's bastard.

During the early days of British sailing, women were allowed to accompany their husbands aboard long voyages. Inevitably, some of these women were not the sailor's wives. Many legitimate, and fewer out-of-wedlock, births took place on ship, and most babies were delivered in a screened-off section of the gun deck. Sometimes the newborn babies actually slept in a hammock attached to the gun barrel. *Son of a gun,* then, probably originally referred to the unknown paternity of a woman's offspring.

Why Are Useless Things *Not Worth a Tinker's Dam?*

Today, it is hard to conjure up a time when there were actually people whose profession was to mend pots and pans. But such was the calling of tinkers.

DAVID FELDMAN

One of their most common tasks was to patch up holes in cooking utensils. Soldering solved the problem, but tinkers had to devise a way to keep the solder from going in one side of the hole and out the other. So tinkers would create a "dam," made of mud or clay, which would keep the solder in place until it had set properly. Once the hole was properly patched, the "dam" was rendered useless and was thrown away. Clay isn't exactly the most glamorous substance to begin with; once it had served its purpose, it *wasn't worth* [even] *a tinker's dam.*

Submitted by John H. Thompson of Glendale, California.

Why Is an Obsequious Sycophant Called a *Toady*?

Toady is a nineteenth-century slang expression for a *toadeater*. Yes, one who "eats" toads.

In the nineteenth century, frogs were considered a delicacy, but toads were widely believed to be poisonous. Traveling medicine shows would often feature an assistant who would pretend to or actually would eat a toad and then suddenly fall victim to a fictitious dread malady. The medicine man, of course, just happened to have the elixir that would cure the unfortunate boy and, marvel of marvels, actually had some of the miraculous potion left over to sell to the public.

The *toadeater*, soon known as a *toady*, was originally somebody willing to endanger or humiliate himself for his master. Today, of course, a *toady* is more likely to go far in the business world than to be a servant, and is more likely to be swallowing his pride than swallowing amphibians.

Why Is an Awkward or Boorish Person Called *Uncouth?*

Sitcom writers know that it is always good for a laugh to have an uneducated character call someone else "couth." But *couth* is a legitimate word and actually has ancient origins. In Old English, *couth* meant "known" or "familiar." *Uncouth,* logically enough, meant "unknown," or "unfamiliar." Because we always have tended to fear and distrust the unknown, *uncouth* developed bad connotations ("awkward" and "boorish") over the centuries.

Couth has recently become a catchword through back-formation. In back-formation, a word appears to be the base of another word when it is actually formed from the other word. Thus, although *burgle* seems to be the base for *burglar, burglar* actually preceded *burgle. Uncouth* is a more complicated case, because *couth* came first in its archaic sense ("unfamiliar"). Then came *uncouth,* meaning "boorish." But the current use of *couth* to mean "class" (as in "You have to admit it, the man's got couth") clearly is a back-formation from *uncouth.*

DAVID FELDMAN

Words About Words

Why Does *That's All She Wrote* Mean "You're Gone, Buster!"?

Because this expression, which came into vogue during and after World War II, refers to the same heartless "she" who wrote the infamous "Dear John" letter.

Submitted by Joan Den of Huntington Beach, California.

WHO PUT THE BUTTER IN BUTTERFLY?

How Did & Come to Mean *And,* and Why Is the Symbol Called an *Ampersand?*

The symbol & was first used as *and* in the twelfth century. It replaced the word *et* (Latin for "and"). The symbol is a combination of the capital "E" and a "T," with the upper loop of the "E" substituting for the cross of the "T."

Well into the nineteenth century, throughout England and much of the United States, the ampersand was taught to children as part of their *ABC*s. It became, in effect, the twenty-seventh letter of the alphabet. Teachers wanted students to learn that *A* and *I* were words as well as letters. They differentiated these letters by teaching the children to recite "*A* per se '*A*' " (meaning "*A* by itself means '*A*' ") and "*I* per se '*I*.' " After reciting the twenty-six letters of the alphabet, children were taught to say "*and* per se *and*" (meaning "*and* by itself means 'and' "), but children slurred the phrase to *ampersand.*

The *Oxford English Dictionary* cites 1837 as the first recorded reference to the word that kids created—*ampersand.*

Why Is Gibberish Called *Gobbledygook?*

How appropriate that *gobbledygook* was coined by a politician! During World War II, Congressman Maury Maverick of Texas, made up the word spontaneously during a speech. He compared the verbiage of a colleague to the turkeys back home in Texas. Not only did both the gobbler and his political enemy spout uninterrupted verbiage, but both strutted with undeserved pretension. As Maverick himself said, "At the end of this gobble there was a sort of gook."

DAVID FELDMAN

Yes, Maury was related to the Maverick immortalized by James Garner in the television series Maverick. The reason that the name of Maury's grandfather, Samuel Johnson Maverick (1803–70), has long stood for "independent free-wheeler" is that old Sam refused to brand his own cattle but claimed any unbranded range stock he came upon as his own.

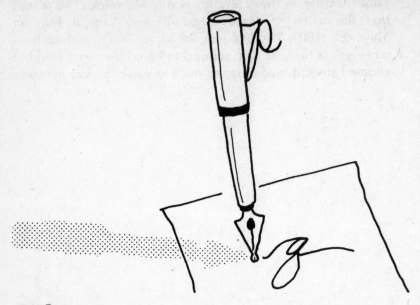

Why Do We *Mind Our P's and Q's* Instead of Our *V's and W's?*

This often-asked question has been researched thoroughly by lexicographers over the years but remains impervious to definitive solution. Three viable theories have been advanced. In descending order of likelihood:

1. *Mind your p & q* ("pints and quarts") was a common alehouse expression in the seventeenth and eighteenth centuries. Barowners would keep a tally on a slate of all ale orders. When a customer was charged for a pint but drank a quart, or when the publican noticed an account was overdrawn, he might chime in with the warning *"Mind your p's and q's."*

2. Printers always have had difficulty setting letters that are inverted or are mirror images (such as *b* and *d* or *p* and *q*). *Mind your p's and q's* might have been a warning to printers to concentrate on the task at hand. Teachers of handwriting warn of *p's* and *q's* for the same reason.

DAVID FELDMAN

3. In the seventeenth and eighteenth centuries, when the minuet was all the rage, young dancers were warned to *mind your pieds and queues* ("feet and pigtails"). The minuet involved bowing, and young men, unaccustomed to wearing wigs, were likely to lose their hairpieces if they weren't careful.

> *Submitted by Mrs. Gloria Greco of Rochester, New York. Thanks also to Joseph Surgenor of North Vancouver, British Columbia.*

What Angry Person First *Read the Riot Act?*

George I of England gave new meaning to the words "law and order." In 1716 he instituted the Riot Act, which made it illegal for twelve or more people to congregate together and "disturb the peace." If the crowd did not disperse, they were subject to a minimum of three years in prison.

George put quite a burden on all public officials. If they encountered such a crowd, they were obligated to stand before the crowd and literally *read the Riot Act,* ticking off the provisions of the law, which undoubtedly must have endeared them to drunken revelers or political protesters.

Why Do We Say That Someone Who Is Talking "A Mile a Minute" Is *Talking a Blue Streak?*

The "blue" reference in this 1830s American expression is to the blue skies, whence "streaks" of lightning emanate. Most of us would rather hear thunderbolts than the cacophony spewed by a human talking a metaphoric *blue streak.*

Why Are Vague Euphemisms or "Bureaucratese" Referred to as *Weasel Words?*

In 1916, during World War I, Woodrow Wilson suggested implementing a policy of "universal voluntary training," which sent Theodore Roosevelt into a fit of fury. Roosevelt saw "universal voluntary training" for exactly what it was—bureaucratic gibberish intended to soften what was actually a proposal for conscription. We could have universal training or we could have voluntary training, but we couldn't have both, and Roosevelt castigated Wilson for his cowardly use of *weasel words,* words that actually rob the sentence they are in of any meaning because they cancel each other out.

The expression *weasel words* was actually coined by writer Stewart Chaplin in 1900, who pointed out the use of *duly,* as in *duly protected,* as "always a convenient weasel word." But why did Chaplin and Roosevelt defame the poor weasel, who, after all, helps rid the world of excess mice and rats?

DAVID FELDMAN

Theodore Roosevelt made the reference clear. A weasel is capable of sucking out the contents of an egg without breaking the shell. Likewise, the "weaseler" is able to rob a sentence (or a whole speech) of its meaning, leaving it empty, while still giving the appearance of utter earnestness and erudition.

By the Numbers

Why Is a Brutal Interrogation Called the *Third Degree?*

The "third degree" is neither a reference to a third-degree burn nor to third-degree murder, even though the phrase conjures up both a criminal interrogation and the possibility of pain and torture. A misunderstanding of the Freemasons is to blame for the frightening connotations stirred up when one contemplates receiving the third degree.

Masons must take examinations before ascending the ranks of the organization. The first and second stages require little in preparation or performance. The third degree (Master Mason) is achieved only after passing a slightly more elaborate test. Because Freemasons were secretive about their customs, rumors circulated that the ritual required to achieve the third degree involved arduous mental gymnastics and brutal physical punishment. Although this allegation was totally unfounded, the Masons' exam was compared to the interrogation and physical badgering of a suspected criminal by the police.

Criminals sometimes refer to their arrest as the *first degree;* the escorted trip to the jail as the *second degree;* and, of course, their questioning as the *third degree.*

Why Does a Drunk Person Have *Three Sheets to the Wind?*

Stuart Berg Flexner points out that whoever first coined this expression was undoubtedly a landlubber who mixed up his terminology. A "sheet" is not a sail but is the rope or chain attached to the lower corner of the sail. By shortening or extending the sheet, one can determine the angle of sail. If one loosens the sheet completely, the sail flaps and careers. If one loosens all three sheets, the ship would reel like a drunk person. Since "in the wind" had long referred to a ship out of control, *three sheets to the wind,* first in print in 1821, was the perfect way to describe the fool who has imbibed two too many.

Why Do We Call the Destruction of a Person or Thing Deep-Sixing?

Deep six is an old expression, originally meaning "a grave."
Why not *deep eight*? Probably six was chosen because of the

custom of digging a grave six feet deep (thus the expression *six feet under*).

Deep six was particularly popular among sailors, and it is likely that the reference was to six fathoms. Sailors used *deep six* to refer specifically to drowning victims (i.e., anyone six fathoms—thirty-six feet—down in the sea was in a literal or metaphorical grave) and also to equipment jettisoned overboard that fell down to the bottom of the sea.

We have John Dean's testimony during the Watergate hearings to thank for the resurgence of this expression: Dean, President Nixon's counsel, testified that when he informed John Ehrlichman that there were incriminating documents found in Howard Hunt's White House safe, Ehrlichman suggested that it might be prudent to deep-six them—in the Potomac River.

If We Are Euphoric, Why Are We on *Cloud Nine?*

The Weather Bureau subdivides each class of clouds into nine types, so some word experts speculate that *cloud nine* refers to the high-flying cumulonimbus clouds that waft upward of twenty-five thousand feet above the earth. The problem with this theory is that several other clouds attain even greater heights. Wouldn't bliss better be described by the cloud closest to the heavens?

Others have speculated that *cloud nine* is a reference to the ninth heaven in Dante's *Paradise* (i.e., next to God, Who resided in the tenth heaven). Neil Ewart conjectures that *cloud nine* is derived from *cloud seven* or *seventh heaven.* Ancient astrologers believed that seven planets governed the fate of the universe and that the seventh was where God resided. Ewart offers no convincing explanation for how cloud seven got inflated by two.

We prefer a less lofty but more concrete explanation for this term. On the popular radio show of the early 1950s, *Johnny Dollar*, the eponymous hero, every episode, knocked himself unconscious. In a gimmick as central to the show as Fibber McGee's closet or Dagwood's sandwich, Johnny was then transported into a blissful place. Upon awakening, Johnny would regale the audience with fanciful reports of what he saw and did in paradise. Johnny called this wonderful place *cloud nine*. Since *cloud nine* did not gain popularity as a catch phrase until the 1950s, it is safe to assume that *Johnny Dollar* was responsible for this contribution to our lexicon.

Why Is a Passing Fad or a Short-Lived Celebrity called a *Nine*-Day Wonder?

In our Warholian times, when anyone can be a celebrity for fifteen minutes, nine days in the limelight constitutes a career. Nine whole days of success? Time to retire and enter the Hall of Fame!

Chances are, *nine* was chosen as the lucky number because a novena in the Roman Catholic Church consists of nine days of special prayers, and many religious festivals of the Church traditionally lasted for nine days. A minority contingent insists that the origin of *nine-day wonder* lay in the proverb "A wonder lasts nine days, and then the puppy's eyes are open."

The cliché certainly has a fine pedigree in literature. Chaucer says in *Troilus and Cressida* (1374), "Eke wonder last but nine daies. . . ." And in Shakespeare's *Henry VI*, Part 3:

> Gloucester: That would be ten days' wonder at the least.
> Clarence: That's a day longer than a wonder lasts.

DAVID FELDMAN

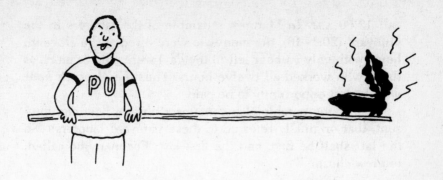

Why Is the Pole That You Won't Touch Somebody or Something with Always *Ten Feet* Long?

In early America, ten-foot poles abounded. They were used in pole boats, flat-bottomed vessels designed to haul farm products or household goods in shallow waters. Poles were essential to navigate through swamps studded with mud bars but also were used in rivers and lakes.

Why ten-foot poles? A long pole enabled the rivermen to push off from the shore or potential impediments and to push up when the pole boat got stuck in mud. The poles, because of their uniform length, became measuring sticks as well, handy devices to ascertain the depth of the water.

Why Do We Call the Last Moment Before an Anticipated Event the *Eleventh Hour*?

In biblical times, the typical workday was twelve hours, and hours were counted from dawn rather than from what we now call 12:00 A.M. In Matthew's parable of the laborers in the vineyard (20:1–16), the men who were hired at the eleventh hour (with only an hour left in the day) were paid as much as those who worked all twelve hours. Thus the *eleventh hour* was the last opportunity to be paid.

The same parable also contains a phrase, now omnipresent, that originally referred to these vineyard laborers: "So the last shall be first, and the first last: For many be called, but few chosen."

Why Is the Only Number You See Before *Skidoo 23?*

Who would have thought that this breezy bit of slang has lofty roots? It does, in Charles Dickens' *A Tale of Two Cities*. The hero of this sad novel is Sidney Carton, who is the twenty-third of a multitude executed by the guillotine.

In the last act of the theatrical adaptation, *The Only Way*, an old woman sits at the foot of the guillotine, calmly counting heads as they are lopped off. The only recognition or dignity afforded Carton as he meets his fate is the old woman emotionlessly saying "twenty-three" as he is beheaded.

"Twenty-three" quickly became a popular catchphrase among the theater community in the early twentieth century, often used to mean, "It's time to leave while the getting is good." Cartoonist T. A. Dorgan combined "twenty-three" with "skidoo." Skidoo was simply a fanciful variant of "skedaddle."

Why Does *86* Mean That a Restaurant Item Is No Longer Available?

Actually *86* has developed alternate meanings as well. A person not to be served alcohol (usually because the customer is already intoxicated) is referred to as an *86*, and the process of ejecting said customer is often called *86ing*. Sometimes *86* is used as a generic term for killing or annihilating something (e.g., "The project wasn't cost-effective, so we 86ed it.").

Some etymologists have speculated that all of the meanings of *86* have a negative connotation because "6" rhymes with "nix." We do know for sure where the expression originated: in soda fountains and lunch counters in the United States during the 1920s.

Before fast-food stores and modern coffee shops, lunch-counter clerks would take orders from customers and, usually without writing a check, yell the order to the cook. To save time and to communicate with customers understanding what they preferred to remain secret, cooks, lunch-counter clerks, and soda jerks developed a shorthand verbal code. The term *86* had two meanings. Most often it signified that the cook was all out of that item. Less frequently it was a code for indicating that a customer should not be served (usually because the patron was a deadbeat).

Most of today's colorful diner lingo developed in the soda fountains and lunch counters of the 1920s. Why say "Please toast an English muffin" when "Burn the British" is available? Only a soulless clod would ask for salt and pepper; waiters with a poetic bent would ask for "Mike and Ike" instead. Occasionally the lunch-counter code was more longhand than shorthand. "Liver and onions" became "Put out the lights and cry."

Soda jerks were big on numbers. Some examples:

19: Banana split

33: Cherry cola

51: Hot chocolate

55: Root beer

80 or *81:* Glass of water

For some of the most popular drinks, such as water and milk, the first digit signified the beverage and the second the number of glasses. Thus *80* or *81* would indicate one glass of water; *82*, two glasses of water; *85*, five glasses of water, etc.

Some of the number codes were veiled warnings from one employee to the other:

13 (obviously an unlucky number): The boss is here. You'd better not mess around.

95: A customer is leaving and trying to stiff us. (A *95* today becomes an *86* tomorrow.)

99: Gulp. The boss wants to see you.

Although remnants of the lunch-counter verbal codes still can be heard in diners and truck stops across the United States, the number code, except for *86*, has largely disappeared. Now that all restaurants use checks and most cooks are away from customer earshot, the need for these colorful terms has vanished.

Submitted by Pat O'Conner of Forest Hills, New York. Thanks also to Sharon M. Burke of Los Altos, California, and to William C. Stone of Dallas, Texas.

Law, Finance, Politics, and Other Disreputable Activities

Why Do Lawyers Call Themselves *Attorneys-at-Law*?

What's the difference between an attorney and a lawyer? Probably about fifty dollars an hour.

Several readers of the *Imponderables* books have pointed out the apparent redundancy in *attorney-at-law*. It's getting harder and harder to find a lawyer who doesn't prefer to be called *attorney*, or even better, the highfalutin *attorney-at-law*.

The roots of the word *attorney* go back to the Indo-European *ter*, meaning "to turn." The later Latin *attorn* meant "to turn over to another." The earliest attorneys, then, were not necessarily lawyers, but anyone designated to take the place of another in a transaction. For example, John Ciardi quotes *A Short Catechism* (1553): "Our everlasting and only High Bishop, our only attorney, only mediator, only peace maker between God and man." This quote clearly marks *attorney* as meaning a mediator rather than simply a lawyer.

Up until around 1800, an attorney was anyone authorized to act for another in legal or financial matters, and to a certain extent we retain this sense today (we need not be a lawyer to retain *power of attorney*). So the phrase *attorney-at-law* was originally a useful distinction, indicating a lawyer able to represent the legal interests of others.

Despite the contemporary interchangeability of *attorney* and *lawyer*, a useful distinction remains. *Attorney* should probably only be used to describe someone who actually represents the interests of a client. A professor who confines his activities to studying and teaching law should be called a *lawyer*. And a pretentious attorney should be called an *attorney-at-law* because it will make him or her happy.

With the proliferation of title inflation in America, I'm surprised that more professions haven't cashed in on the original meaning of *attorney*. Imagine shingles laden with *attorney-at-accounting* or *attorney-at-real estate*.

Submitted by Kevin T. Jones of Baytown, Texas.

Why Are Constantly Fighting People Said to Be *at Loggerheads?*

In the warm waters of the Atlantic Ocean, one can find snapping turtles called *loggerheads*, so named because they have big, knobby heads. But they are not responsible for this phrase. In fact, the turtles are named after the same weapons that generated *at loggerheads*.

Medieval navies carried long-handled sticks with a solid ball of iron at the end. The iron end was heated and used to melt tar, which was then flung at the enemy. If the tar supply

DAVID FELDMAN

was depleted, the combatants would bash each other with the sticks.

At *loggerheads* was recorded as early as 1680 in England. Its original meaning was "obstinate or ignorant blockhead," a clearly pejorative sentiment that lives on today. People who are *at loggerheads* are not only feuding but also are unable and unwilling to compromise or to seek a reasonable solution to their problems.

Why Are Optimists in the Stock Market Called *Bulls* and Pessimists Called *Bears?*

Yes, when they attack, bears sweep their paws downward and bulls thrust their horns upward, but despite folk wisdom, this isn't how these expressions originated. *Bears* was first recorded in 1709, a full five years before *bulls*.

DAVID FELDMAN

The original *bears* would now be called *short sellers*. Believing that prices would fall, speculators sold shares they didn't own, anticipating that they could buy the stocks later at a low price. An Old English proverb, "Don't sell the bearskin before the bear is caught," obviously applied to the risky strategy of the speculators, so they were called *bearskin jobbers*, and later, simply *bears*.

Optimists who bought stocks in anticipation of a price rise probably were called *bulls* because of the alliteration with *bears* and because the two animals were already linked in a then popular sport, "bear- and bull-baiting."

Why Does *Buck* Mean " a Dollar"?

Buck has meant "male deer" since the year 1000 in England and has meant "a dollar" in America since 1856. Despite the time gap, the two meanings are closely linked. In the early eighteenth century, traders and hunters used buckskin as a basic unit of trade. Any frontiersman who possessed many buckskins was considered a wealthy man.

How did *buck* come to mean specifically *one* dollar? In the early West, poker was the diversion of choice. A marker or counter was placed to the left of the dealer to indicate who was the next to deal. This marker was traditionally called *the buck*, because the first markers were buckhorn knives. But in the Old West, silver dollars (i.e., one dollar), instead of knives were used as bucks.

The *buck* as poker counter yields the expression *pass the buck*, a favorite of politicians and bureaucrats everywhere, who usually are more than happy to evade responsibility for governing, dealing poker, or just about anything else, which was why it was so surprising to hear Harry Truman, an admitted poker player, announce, "The buck stops here."

Why Is the Death Penalty Called *Capital* Punishment?

The *capital* in *capital punishment* has nothing to do with the seat of government. The word derives from the Latin *caput*, meaning "head." The original punishment for a capital crime was the loss of one's head. The reason the first letter of a sentence is said to be *capitalized* is because it is at the "head" of the sentence.

Why Is a Police Officer Called a *Cop*?

If we know when a word or expression was first used, we have the first important clue to how, where, and why it came into the language. Many stories have circulated about the origins of *cop*. A common explanation is that *cop* is an acronym for constable *on patrol*. Another is that *cop* is short for *copper*, the metal on the buttons worn by London's bobbies. The only problem with either theory is that the use of the word *cop* predates either of the above uses. *Bobbies* didn't exist until 1829, but *cop* was recorded in print as early as 1704.

Cop almost certainly derives from the Latin *capere* ("to capture"), which was transformed into the French *cap* and eventually the Middle English *cop*. Since a police officer's job is to capture criminals, *cop* is an appropriate nickname.

Our *cop*, then, is simply the noun form of the verb we use all the time to mean "capture" or "take." A furtive teenager *cops a feel*. And a gangster, whose subculture spread the use of the word *cop* in the first place, *cops a plea* (originally, "to confess to the cops").

Submitted by Venia Stanley of Albuquerque, New Mexico.

DAVID FELDMAN

Why Is Someone Who Surreptitiously Listens to Others' Conversations Called an *Eavesdropper*?

Eavesdropping isn't exactly an endearing activity today, but from the sixteenth through the nineteenth centuries, eavesdropping was a crime in England. Back then, communities were not equipped with gutter systems, so houses were surrounded by eavesdrops, spaces all around a building where water dripped from the eaves. The purpose of the eavesdrop was to allow a wide overhang so that rain fell far enough from the house to safeguard the security of the foundation.

The first *eavesdroppers* were nefarious types who literally stood in the eavesdrops to overhear private conversations. Protected from the elements by the overhang, this low-tech espionage evidently faded as sewer systems rendered eavesdrops obsolete.

Submitted by Patricia Fox-Sheinwold of New York, New York.

Why Do We Call a Politician's Verbal Delaying Tactics a *Filibuster?*

Our term *filibuster* comes directly from two Dutch words: *vrij* ("free") and *buit* ("boot"). These two words, in English translation, yielded the word *freebooter* in the sixteenth century.

The French appropriated the Dutch words and transformed them into *filibustier;* the Spanish modified it to *filibustero*, the immediate parent of our *filibuster*. The French and Spanish versions meant the same thing as *freebooter*—"a pirate."

But in the United States, *filibuster* was first used to describe adventurers who fomented revolution in the Spanish colonies of Central American and the Caribbean. These adventurers were reputed to harangue their listeners with partisan rhetoric. Was it not an appropriate analogy to so describe legislators who avoid action (or even worthwhile debate) by prolonged, blustery monologues?

DAVID FELDMAN

Why Do We Call a Five-Dollar Bill a *Fin?*

No self-respecting gangster, let alone a heavyweight fictional one like an Edward G. Robinson character in a 1930s movie, would be caught dead saying "five dollars" when *fin* was available. *Fin* was first recorded in 1929, just in time for those wonderful 1930s gangster movies, and seems to have sprung genuinely from the lingo of the underworld.

Fin, short for *finif*, has Yiddish origins and means "five." Since the late eighteenth century, *fin* had been used as slang for "hand," and the fact that there are five fingers on the human hand served as a reminder to the straight-and-narrow population to which bill *fin* referred.

Submitted by Tom and Marcia Bova of Rochester, New York. Thanks also to Robert J. Abrams of Boston, Massachusetts.

Why Is Redistricting to Gain Political Advantage Called *Gerrymandering?*

Elbridge Gerry was one of the signers of the Declaration of Independence. A rabid Jeffersonian and anti-Federalist, Gerry, governor of Massachusetts in the early nineteenth century, was not beyond being party to a little questionable manipulation to realize his goals.

Gerry had a large map of Massachusetts on his wall. For the 1812 election, Essex County was redistricted by Gerry's party to favor Jeffersonian candidates. The resultant district was so contorted that someone (various accounts name newspaper editor Benjamin Russell or painter Gilbert Stuart) likened the appearance of the new district on the map to a salamander. Gilbert, supposedly, added a wing, claws, and a head to make the outline of Essex County look even more like a salamander on Gerry's personal map.

Gerrymander, then, is the combination and elision of *Gerry* and *salamander*. Ironically, the gerrymandering was of no help to Gerry himself. He certainly couldn't find a way to gerrymander Jeffersonians from neighboring states into Massachusetts—he lost in his reelection bid for governor. Gerry was rescued from obscurity, however, by becoming James Madison's vice president during his second term. Gerry died while serving as vice president, in 1814.

Gerry's name was pronounced with a hard "G," so although most Americans pronounce it *"Jerry"mander*, *"Gary"mander* has the original pedigree.

Why Is an Angry Uproar or Mob Resistance Called a *Hue and Cry*?

The *cry* comes from the French *crier* ("to cry out"). The *hue* has nothing to do with color but with another French word, *huer* ("to shout"). In England in the Middle Ages, being a good Samaritan was not voluntary. If one heard the scream, *hu e cri*, it meant either that someone was being victimized or that an officer needed help apprehending a criminal. In either case, if a bystander didn't help solve the problem by offering his assistance, he was subject to punishment. *Hue and cry* statutes existed for hundreds of years; in fact, the last statute was repealed as late as 1827.

DAVID FELDMAN

Why Do We Call Seizing a Child or an Adult *Kidnapping?*

Etymologically, *kidnapping* has a mundane story. *Kid* has meant "child" since ancient times, and *nap* was a seventeenth-century variation of *nab*.

But the historical tale behind the origins of *kidnapping* is a grim reminder about early American history. The early colonies in the New World were in desperate need of unskilled farm labor and apprentices to skilled craftsmen. There simply weren't enough bodies available in Jamestown and other colonies, so the only way to recruit needed labor was to lure the downtrodden in England.

The colonists offered Britishers the enticement of indentured servitude. The owner received seven years of free labor from the servant. In exchange, the servant received free passage to the New World (although not in *QE II*-type accommodations), and a modest payment after the seven years of service, usually in the form of clothing and tools of the servant's trade.

Indentured servitude worked relatively smoothly in the seventeenth century until the colonists ran out of volunteers. Americans had used professional recruiters in England, who were paid on a commission basis to lure away able-bodied laborers. These recruiters were called *spirits*, and they sold the soon-to-be servants to ship captains. The spirits, then, were sort of human wholesalers, never directly in touch with their New World customers. Indentured servants were worth about five pounds sterling in the early eighteenth century, and the spirit, sharing with the ship captain, would receive only a fraction of this price.

As potential recruits increasingly expressed the sentiment that indentured servitude was an offer they *could* refuse, spirits became desperate to gather "merchandise." Bereft of volunteers, unscrupulous spirits *spirited away* (yes, that is the genesis of this expression) innocent and often unwilling children. *Kidnapping* originally referred specifically to the abduction of children by spirits.

Kidnapping became such a serious problem that in 1682 the London Council passed a law specifying that no child under fourteen years of age could be indentured without the consent of his parents.

Submitted by Phyllis Diamond of Cherry Valley, California.

DAVID FELDMAN

Why Is a Woman's Allowance or Personal Expense Reserve Called *Pin Money?*

We tend to think of pins, staples, and paper clips as natural resources, found objects. But the pin wasn't invented until the sixteenth century in France, and they were hardly throwaway items. Early pins were made of silver and thus very expensive. In England, a monopoly under grant from the crown had exclusive right to manufacture pins, and they chose to produce a small amount and charge dearly for them.

Henry VIII's fifth wife, Catherine Howard (1520–42), popularized this French invention, and her subjects were clamoring for pins. According to Neil Ewart, the general public could buy pins only once a year, during two days at the beginning of January. Husbands gave their wives money to buy the pins; thus the origin of *pin money* as a private stash of cash for women.

Eventually, pin money became an inherent right of wives. In divorce suits in England, women often sued to collect one year's pin money, as much as two-hundred pounds a year. Pins are one of the few household items that have undergone great deflation over the years.

Why Is the *Left Wing* Liberal and the *Right Wing* Conservative?

The world wide epithets for "liberal" and "conservative" come from the European tradition of legislators placing conservative parties to the right of the chair and liberals to the left.

Perhaps the seating was originally random, but we suspect that since right-wingers tend to come from a higher social background than their more radical counterparts, they were given the favored position. Just about everything *right* in our culture is "good," and everything *left*, if not bad, is a little strange. The aristocrats of the conservative parties got the "better" seats, just as they tend to in restaurants these days.

DAVID FELDMAN

Why Does *Lucre* Always Seem to Be *Filthy*?

Lucre is not merely a synonym for *money*; it is tainted money, loot, obtained by dishonorable deeds or intentions. Many languages have an equivalent word. Ours is derived from the Latin *lucrum*, which means money without any negative connotations. But Sanskrit has *lotra* (stolen goods) and Hindu *lut* (loot).

Although most of us associate "filthy lucre" with the dialogue in bad Westerns, we borrowed the phrase from the Bible. In Timothy, Paul describes the qualities necessary in a bishop: "Not given to wine, no striker, not greedy of filthy lucre." Paul also warns of deceivers who are "teaching things which they ought not, for filthy lucre's sake." Christ drove the money changers from the temple, demonstrating his disdain of economic exploitation. Many other religions have strictures against usury.

In *A Browser's Dictionary*, John Ciardi notes that from the inception of money, most cultures held strong convictions that "true value consisted of land, crops, and livestock, and that minted money upset a natural balance. Money was proverbially said to be "the root of all evil."

With our strong cultural heritage of ambiguity about money, no wonder the word "filthy" was chosen to modify tainted money. If any readers would like to rid themselves of the pestilence of *filthy lucre*, please send said filth to the address at the end of the book.

Why Is a Murder as a Form of Vigilante Justice Called a *Lynching*?

Captain William Lynch, of Pittsylvania County, Virginia, author of the infamous lynch laws, will forever be linked with vigilante justice. Lynch felt that he and his neighbors were too far from lawmakers and sheriffs to punish properly the vandals and robbers terrorizing his rural area. He encouraged his fellow citizens to sign a declaration he drafted, announcing the intention "to take matters into their own hands": "If they [criminals] will not desist from their evil practices, we will inflict such corporeal punishment on him or them, as to us shall seem adequate to the crime committed or the damage sustained."

Although the death penalty was not always exacted, in most cases the punishment turned out to be hanging. A certain amount of doubt and guilt among the lynchers can be ascertained by Captain Lynch's technique for hanging criminals. Rather than stringing up the condemned on trees (the preferred method of subsequent southern *lynchers*), Lynch and his cohorts practiced a form of passive hanging. A rope was tied around a tree and the condemned man placed on a horse with the other side of the rope strung snugly around his neck. The criminal was killed not by his captors tightening the noose but by the whim of the horse. When the horse moved far enough away from the tree, the rope choked the neckless horseman.

DAVID FELDMAN

Why Is an Independent Person, Especially a Great Person, Called a *Mugwump?*

Mugwump is one of those unusual words that can be used earnestly or facetiously. Some would love to be called *mugwumps;* others would prefer so calling their enemies. The history of the word is equally ambiguous.

The original mugwumps were a splinter group of the Republican Party in 1884. Refusing to support the presidential nominee of their party, James G. Blaine, mugwumps supported the Democratic Party's nominee, Grover Cleveland. They were dubbed *mugwumps* by their Republican rivals. *Mugwump* means "chief" in the Algonquin language, and the word spread among non-Indians, especially in Massachusetts. Obviously, the original non-Indian use of *mugwump* was sarcastic. The party regulars were accusing the dissenters of self-righteousness and delusions of greatness.

Yet, the mugwumps themselves took the epithet as a badge of honor, a term indicating their integrity and independence. Others have used *mugwump* to designate a turncoat. But in its narrow, more precise definition ("one who stays within the arms of one party but doesn't vote for the party's candidate"), a mugwump is unlikely to make anybody very happy and is likely to be the butt of ridicule. As Albert Engel defined a *mugwump* in 1936, "A *mugwump* has his *mug* on one side of the political fence and his *wump* on the other."

Why Is a Ten-Dollar Bill Called a *Sawbuck?*

A sawbuck is a kind of sawhorse, one whose legs form a cross pattern as they project above the crossbar. The pattern of the four protruding crossbars looks like an *X. X*, of course, is the Roman numeral designation for ten, and since the Roman numeral *X* appeared on ten-dollar bills at the time when *sawbuck* was first recorded (1850), it isn't quite as strange as it first appears that we should get our nickname for a ten-dollar bill from a carpenter's prop.

Why Do We Say That Someone Who Has Appropriated Someone Else's Ideas or Future Remarks Has *Stolen Thunder* from the Victim?

I had always assumed that this expression must have Greek mythological roots and perhaps was a reference to Zeus. The true story is much more prosaic.

John Dennis, an English poet and playwright, wrote a tragedy called *Appius and Virginia*, which was produced in 1709 to less than rousing commercial success. Only one element of the production stirred the audience: thunder sound effects more realistic than any heard before on the stage, effects that Dennis himself created.

The play failed, but the theater's next production didn't. Dennis went to check out a successful production of *Macbeth* and was more than a little upset to discover that his sound effects were used in the storm scenes of Shakespeare's tragedy.

Different sources vary slightly in describing what Dennis exclaimed upon hearing "his" thunder help promote the new production, but they are all variations of Stuart Berg Flexner's quote: "See how the rascals use me! They will not let my play run, and yet they steal my thunder!" I'm sure that Dennis would be even more embittered to learn that the only phrase of his that has gained immortality is his expression of sour grapes.

DAVID FELDMAN

Clothes Make the Word

Why Does *Hoodwink* Mean "to Fool" or "to Blindside"?

The original meaning of *wink* was "to close one's eyes," with no implication that the closing was voluntary or a kind of signal to another. In the sixteenth century, when cloaks were a fashion must, cowls or hoods were attached to the cloaks.

When one was hoodwinked, then, one was literally blinded by the hood. Cognizant of the phenomenon, industrious thieves preyed upon unsuspecting suckers whose peripheral vision was extremely limited.

Why Is a Nude Person Said to Be Wearing His *Birthday Suit?*

Although some etymologists insist that this expression stems from a custom of English kings of buying a new suit of clothes for his retinue on the royal birthday, we can't agree. Why would an actual suit come to mean "no clothing at all"? More likely, the explanation for the genesis of *birthday suit* is considerably more prosaic: It describes exactly what one was wearing on the *day* of one's *birth.*

Why Is a Riding Costume Called a *Habit*?

Habit is a Middle English word. Originally it referred to any dress or costume, then to any dress worn "habitually," such as a uniform. *Habit* was also applied to certain costumes worn only for certain occasions, such as a riding *habit* or a fencer's suit. All of these meanings precede our primary contemporary use of *habit* as something done automatically and often involuntarily.

The word "custom" had a similar history. Originally *custom* meant "costume" but now has much broader applications.

Who Was the Jean That Blue *Jeans* Are Named After?

This early-sixteenth-century word, which originally described the cotton material rather than the garment it made, was not named after a person but after a city. *Jean* is a derivative of *Genoa*, Italy.

Denim similarly derives from the city of Nîmes, France. The material originally was called *serge de Nîmes*.

And yes, there really was a Levi Strauss. Strauss, a San Francisco merchant during the Gold Rush days, added rivets to the corners of the pockets, making *Levis* a handy as well as durable pair of pants.

How Did the *Pea* Jacket Get Its Name?

The *pea* in *pea jacket* has nothing to do with the vegetable or the color of the vegetable. Nor is *pea* short for the *p* in *pilot*. In fact, the first pea jackets were not pilot jackets at all, but short, double-breasted coats worn by Dutch sailors in the fifteenth century. As they were made of coarse cloth, they were ideal for use in rough weather.

So why the *pea*? The Dutch name for the jacket was *pijekker* ("coarse cloth"), and when the English adopted the style, they shortened the Dutch word to a more manageable size.

How Did Those Beltless Hawaiian Dresses Ever Get Named *Muumuus?*

Muumuus were created not by Sergio Valente or Bill Blass but by Christian missionaries sent to Hawaii who were horrified by the natives' tendency to parade around unashamedly in the altogether. Despite the way they make the average woman appear, *muumuu* has nothing to do with cows whatsoever. *Muumuu* means "cut off" in Hawaiian, so the name seems to stem from the dresses' lack of a yoke; a muumuu is so shapeless that it seems arbitrarily "cut off" at the neck. Now that it is more than thirty years since muumuus were all the rage, brace yourself: Muumuus are likely to reappear before Halley's Comet.

PFFT!

HOW Did *Sideburns* Get Their Name?

Sideburns are named after General Ambrose Everett Burnside, a man of undoubted charisma but questionable staying power, whose lasting contribution to our culture was lending Elvis a good grooming gimmick and providing Civil War buffs with something to argue about.

Burnside first made his mark by inventing and manufacturing the breech-loading rifle, which he called the Burnside

DAVID FELDMAN

Carbine Rifle. The weapon worked, but the business venture failed. Undaunted, Burnside joined the Union effort during the Civil War, quickly rising to the rank of general.

Unfortunately, Burnside presided over two of the Union's most crushing defeats in the War, at Fredericksburg and Petersburg. Critics contend that his charges were ill-planned and hasty, leaving his forces vulnerable. Burnside was removed from his command.

A failure at business and in the military, what was left for the poor guy? You guessed it: politics! Burnside became governor of Rhode Island and later a U.S. senator. After the war, Burnside, a bit of a dandy, sported long growths of thick black whiskers on his face, what we would now call "muttonchops," and they became a fashion fad in his time.

Why were they called *sideburns* rather than *burnsides*? Actually, they *were* called *burnsides* at first, but by the twentieth century were uniformly called *sideburns*. We have yet to find a convincing explanation for the change and can only speculate that the fact that *burnsides* appear on the side of the face must have influenced the changeover.

Why Is Somebody Obviously in Love *Wearing His Heart on His Sleeve?*

Two charming, outdated customs are responsible for this phrase. In the days of chivalry, knights would wear the scarves, kerchiefs, or favors of female admirers on their sleeves. Since this gesture indicated that the knight reciprocated her affection, he was said to be wearing his heart (i.e., his true feelings) on his sleeve. Some historians are far from sure that this practice ever existed, but it is often portrayed in legend.

We do know for sure that in sixteenth-century England, Valentines were exchanged. If a man was truly smitten by a woman, he wore the heart-shaped Valentine of his beloved on his sleeve, which presumably was more convenient than wearing it on his femur or tibia.

DAVID FELDMAN

How Did the *Zipper* Get Its Name?

Zip was used as a noun and verb in English as early as 1850 and seems to have an echoic origin. *Zip* was probably first used to describe the hissing sound of a speeding bullet. *Zipper* was similarly taken from the sound made by the fastener, and the word was trademarked by B. F. Goodrich in 1925. Zippers helped make Goodrich's overshoes waterproof. Soon the technology was spread to many other applications, especially in clothing. The zipper became so popular that it became a generic term, and Goodrich lost the trademark on the word except for its zippered boots.

Submitted by William Debuvitz of Bernardsville, New Jersey.

Fun and Games

Why Do Weight Lifters Hoist *Dumbbells?*

We could suggest that no smart bell would want to be lifted by a weight lifter, but we are too dignified to engage in this kind of cheap wordplay. And besides, the actual story is fascinating.

The original dumbbells were not free-standing, but were rather complicated mechanisms. The exerciser pulled on a wooden bar attached to a rope that hoisted a weight, similar in theory to our weight machines of today.

These dumbbell machines were designed to simulate the movements of bell ringers, who attained tremendous upper-body development simply by doing their job. To reinforce the idea that one could possess the body of a bell ringer, the original weights were in the shape of a bell, thus accounting for half of *dumbbell*.

In Old English, *dumb* meant "mute" (thus the derivation of *deaf and dumb*, which did not originally have any negative connotations; it meant, simply, "deaf-mute"). *Dumbbells* acquired their name because there were no real bells on the device and thus it made no sound. In essence, the dumbbell was promoted by proclaiming, "Attain all of the body-building benefits of bell ringing without the accompanying earache."

The stereotype of the typical bodybuilder is "all body, no brain," but etymologically there is no connection between *dumb* meaning "stupid" and *dumbbells*, which were introduced in the 1850s.

Is a *Caddy* a "Little Cad"?

Nope. *Caddy* is a corruption of the French *cadet*, the same word now applied to military students. The word was brought into Scotland by Mary Queen of Scots (who was born in France) and originally referred to an errand boy. In golf, the caddy is, in essence, an errand boy.

In France, *cadet* means "younger son." In Mary's time, patrimony was the rule. The oldest son inherited the fortune of the father. Younger sons, even in wealthy families, often had to fend for themselves financially. So gentlemen often volunteered for military service only because they didn't have the good fortune to be born before their siblings.

DAVID FELDMAN

Why Are Two Consecutive Baseball Games Played in One Day Called a *Doubleheader*?

Doubleheader is a straight "steal" from railroad jargon of the late nineteenth century. A *doubleheader* to them was a train with two engines.

Submitted by Douglas Watkins, Jr., of Hayward, California.

Why Do We Call the End Position in Chess *Checkmate*?

Checkmate has nothing to do with either bank drafts or lovers. It comes from a Persian word, *sháh-mát*, meaning "the king is dead."

Submitted by Robert J. Abrams of Boston, Massachusetts.

Why Is the Dice Game Called *Craps*?

The dubious distinction of introducing this game of chance goes to a Frenchman named Bernard Marigny. He brought the game, which was a simplified version of the French dice game *hazard*, to New Orleans around 1800.

Two theories have been advanced to explain the derivation of *craps*, and both are so definitive-*sounding* that one suspects the veracity of either story.

Theory one: Marigny was known as Johnny *Crapaud* ("toad" in French—*toad* was a common nickname for the French in the United States). *Craps*, then, is simply a shortening of *Crapaud*.

Theory two: In the game of *hazard*, the lowest roll (two "ones" or "snake eyes") was known as *craps*. If true, this explains why *craps* is not only the name of the game but also the term for an initial throw of two. Left unexplained, however, is why a roll of three or twelve is also called *craps*, and also signifies an instant loss for the unfortunate thrower.

Why Do We Say That Someone Who Is Taking Careful or Deadly Aim Is *Drawing a Bead On* . . . ?

The bead referred to is not a bauble, but the little knob at the end of the gun barrel used for sighting. Also known as the *foresight* or *muzzle sight*, the *bead* on guns with two sights is always the one closer to the muzzle.

DAVID FELDMAN

We Know What a King Is and What a Queen Is. We Even Know What a Joker Is. But What Is a *Jack*?

Etymologically, *jack* derives from the Old English *chafa* ("boy, male child"). But the clue to the identity of the jack is its alternate name, *knave*. A knave was a servant of royalty, so it is no accident that in a deck of cards, the jack is the lowest-ranked picture card.

Later, *knave* accumulated negative connotations, including "deceitfulness." But knavery usually is associated with rascality, which explains, perhaps, why the jack is often wild in poker games.

Submitted by Mark Anderson of Seattle, Washington.

Why Is a Loyal Partisan Called a *Fan?*

Fan is short for *fanatic*, which was derived from the Latin *fanaticus*, meaning "belonging to a temple" but also meant "a frantic participation in orgiastic rites." Although the earliest applications of *fanaticus* must have been religious, the first recorded English use of *fanatic* referred to a lunatic who no doubt acted somewhat like someone possessed with religious fervor.

Fan first appeared in print in 1896, describing rowdy boxing enthusiasts. By the mid-1920s, *fan mail* became a symbol of a celebrity's fame and popularity.

Why Do We Say That a Person or Team That Scores for the First Time Is *Off the Schneider?*

Schneider is a Yiddish term for a cloth cutter. To the many first-generation Jews in America, cloth-cutting was the entry-level position into the garment industry. Although there was nothing ignoble about the task, it was the lowest rung in the trade; if you were a cloth cutter, you were a nothing, "a zero."

The game of choice among the occupants of the higher rungs of the garment industry was gin rummy. The first time a player scored, he was said to be *off the schneider* (i.e., no longer "a zero"). Rummy terminology later spread to all sports.

Was There a Real Duke Who *Put Up His Dukes?*

The oldest son of King George III was the king of England. The second son was a jerk with a lofty title: duke of York. Frederick Augustus' military career ended in 1809 after he fought a scandalous duel with the Duke of Richmond.

The Duke of York further scandalized English society by pursuing an interest in boxing, which, in the days before Marquis of Queensbury rules, was downright disreputable. York's fascination with boxing gained such notoriety that boxers nicknamed their fists *dukes of York*, and by the end of the nineteenth century the phrase was shortened to *dukes*.

Why Is the Offensive Field General of a Football Team Called a *Quarterback*?

American offensive football formations were quite different in the late nineteenth century. The center was flanked by three linemen on each side, and the *fullback* was aligned behind the center, but quite far away. The two *halfbacks* stood, logically enough, halfway between the line and the fullback. The *quarterback* was actually not directly behind the center, but a little off to either side. The *quarterback* stood only one-quarter the distance behind the line as the *full*back, and thus the derivation of the term.

DAVID FELDMAN

In Golf, Why Does *Par* Mean the "Standard of Excellence"? Why Does a *Birdie* Mean "One Under Par"? Why Does an *Eagle* Mean "Two Under Par"? Why Does a *Bogey* Mean "One Over Par"?

Although we might assume that golfers always had a target to shoot for, *par* (in Latin, "equal") did not exist as a golf term until 1911. *Par* represents how an excellent golfer would expect to score on a given hole. When a golf club committee determines par, the golfer is always allotted exactly two putts on each green. In the United States, par is determined solely by the length of the hole. For men, a par three is any hole less than 250 yards; a par four, 251 to 470 yards; any hole more than 471 yards is rated a par five. There are no par six holes in the United States for men.

Women, who generally cannot hit golf balls as far as men, have a more generous par allotment. Any hole less than 210 yards is a par three; 211–400 yards is a par four; 401–575 yards is a par 5; any hole longer than 576 yards is a par six. Because some country club committees frown on a par six for women, they solve the problem by having women start at the *ladies' tees*, often 20 to 40 yards closer to the hole than the *men's tees*. In the United Kingdom and Ireland, where weather conditions often are severe and their roughs much more untameable, par is not determined solely by the length of the hole, and is up to the discretion of the course's committee.

Birdie, meaning "one under par," is an American invention and a variation of an already accepted slang expression. *Bird* had long meant "excellent" when the Atlantic City Country Club, in 1921, coined this new use of the word. An *eagle*, presumably, was so named because it was a "big birdie." A *double eagle* is "three under par" and rarer than holes in one. Barring flukes, the only chance that most golfers have to achieve a double eagle is to score a two on a par five, a feat considerably harder than making a hole in one on a par three.

Bogey is derived from the Middle English *bugge*, meaning "demon" or "goblin." Although a professional golfer might consider a bogey a devilish curse, many duffers would embrace a bogey as a gift from heaven.

Until 1890, golfers didn't have pars to shoot for. But Hugh Rotherham, a member of England's Coventry Golf Club, created *Rotherham's game*, in which a target was assigned for each hole. To win, the golfer had to equal or better the goal after adding his handicap.

In 1891 the song "Bogey Man" became a smash in England. British players created a mythical creature, Colonel Bogey, who was the demon that golfers tried to beat in Rotherham's game; and a *bogey*, in England, became the target score for a hole. In England, the *bogey* was not the expectation of the great golfer, but a target for the "decent" golfer.

When the rubber golf ball was invented in the United States and gained popularity throughout the golfing world, the English were left with a dilemma. The new balls traveled much farther than the old gutta-percha balls, making the traditional British bogey much too easy to beat. The solution: The British retained the name *bogey* but simply reduced the goal by one stroke per hole. The *bogey*, once the target of the decent golfer, became, like the American *par*, the goal of the excellent golfer. Only decades later did the British, who invented the game, succumb to the American scoring system.

Submitted by Joseph Surgenor of North Vancouver, British Columbia.

Why Do We Say That Someone's Ranking in a Tournament Is His *Seed?*

In most tournaments, the top players are placed in the early rounds so that they don't face each other. If a tournament has sixteen players, the number one seed would face number sixteen (the last-ranked), number two would play number fifteen, etc.

Seeding serves the intended purpose of maintaining the integrity of the later stages of the tournament. But because a low-ranked player must defeat a top seed to advance to the later rounds, seeding also tends to perpetuate the success of the top seeds, since the favorites are playing the worst players in the tournament first.

One suspects that seeding always will exist in professional contests; it is one way for sponsors to recoup their expenses. Spectators pay money to see the top seeds and are disappointed when players they are not familiar with reach the finals of a tournament.

How did the name *seed* develop to describe the top-ranked players? The name refers to the metaphor that the top players, like the seeds of a crop about to be planted, are scattered throughout the different playing brackets.

There's No Place Like Other Places

Why Is Someone Who Escapes Penalty or Punishment Said to Have Gotten Off *Scot-Free?*

Although we are not above dabbling in ethnic stereotypes, this expression has nothing whatsoever to do with Scots' supposed propensity for getting something for nothing (or as close to nothing as possible). *Scot-free* isn't even Scottish, but an Old English expression that, according to the *OED*, originally meant "an amount one owed for entertainment." If one went out with friends, your share of the food and drink would be your *sceot*.

Eventually, the meaning of *sceot* was broadened to include taxes, particularly a tax paid to a local sheriff. So the earliest beneficiaries of being scot-free were citizens who were exempt from paying local taxes or who circumvented paying them.

Why Is *Scotland Yard* in England?

The original Scotland Yard, established by Robert Peel (see "Was the London *Bobbie* Named After a Real Bobby?") in 1829, was placed on the site of the former palace where Scottish kings and queens resided when visiting England to conduct affairs of state or to pay tribute to English royalty. *Scotland Yard* became known as the name of the street as well as the palace.

Although the Criminal Investigation Department of the Metropolitan [London] Police later moved to the Thames Embankment and then to the Victoria area of London, it still retains the name of its original site.

Submitted by Meg Smith of Claremont, California.

DAVID FELDMAN

Dave, Are There Any Great Euphemisms Involving African Countries?

O.K., I admit it. Nobody has ever asked or ever will ask me this question. But indulge the artificiality of the question, because it yields a wonderful answer.

In their delightful compendium *Kind Words: A Thesaurus of Euphemisms*, Judith S. Neaman and Carole G. Silver list many euphemisms for the sexual act. The most bizarre, without doubt, is an English euphemism, *discussing Uganda.*

I assume this phrase started with some skeptical father yelling out to his daughter and her suitor ensconced in the car parked in front of the family house:

"Hey, what are you doing in the back seat?"

"Oh, Daddy," replies the dutiful daughter breathlessly, "We're just discussing Uganda."

Why Is a Late Patch of Warm Weather Called *Indian Summer?*

Much as anything French seems to have sexual connotations, most phrases starting with *Indian* mean "false." Early American colonists had the habit of naming "Indian" anything they found in America that resembled what they knew from home but that differed in some way. *Indian pudding*, *Indian corn*, and *Indian tea* were all originally pejorative terms, meaning "bogus pudding," "bogus corn," and "bogus tea."

Colonists were so paranoid about Indians defrauding them that they even "accused" Native Americans of perpetrating a "false summer," *Indian summer.*

Of course, *Indian summer* is hardly exclusive to the United States. The British have an equivalent expression, *St. Martin's summer,* a reference to St. Martin's Day (November 11), when a late "summer" day might occur.

146

Why Do We Call the South *Dixie?* And How About *Dixieland?*

Many arguments have ensued over the origin of *Dixie*, but one thing is certain: The term *Dixie* was popularized at the same time that one of the most popular black face minstrel performers, Dan Emmett, wrote and performed "I Wish I Was in Dixie's Land," in 1859. Although introduced in New York, the song took the entire country by storm. But when the Civil War erupted, the South proudly adopted "Dixie" as its marching song and unofficial anthem.

The real question is whether Emmett himself coined the term or if he borrowed a word already in circulation. Where could he have found *Dixie?* One theory is that *Dixie* is slang for Jeremiah Dixon, one of the two English surveyors (the other, Charles Mason), who were hired almost a century before the Civil War to resolve the border dispute between Maryland and Pennsylvania. Before the Civil War, the border between the two states was considered to be the boundary between slave and nonslave states. The Mason-Dixon Line, then, had much to do with defining the identity of the South, and the eventual loyalties of the states during the Civil War. Although this theory is plausible, it doesn't explain why Dixon was chosen instead of Mason, or why a name associated with a border between two places should be used to identify a region lying entirely on one side of the border.

Perhaps *Dixie* came from the French word *dix* ("ten"). Before the Civil War, ten-dollar notes were issued by a bank in Louisiana that had *dix*, rather than *ten*, printed for the numeral. How one bank became emblematic of the whole South is problematical, however. Although other theories have been advanced, including one that *Dixie* was an eponym for a lovable northern slave-owner, we probably will never arrive at a definitive answer. Surely, it is possible that Emmett simply liked the euphonics of *Dixie*—it wouldn't be the first time

WHO PUT THE BUTTER IN BUTTERFLY?

that a songwriter chose a title for its sound rather than its meaning.

Dixieland got its name because it was the type of jazz played in New Orleans. Louisiana, of course, was soundly in the Dixie camp during the Civil War.

Submitted by Malinda Fillingion of Savannah, Georgia.

Why Are People from the Backwoods Called *Hillbillies?*

Hillbilly is an American expression first recorded in 1904 to describe rustic hill dwellers in the South, so the "hill" in *hillbilly* is easy enough to understand. But why "billy"? Couldn't we be more respectful and call them *hillwilliams?*

"Billy" had long meant "fellow" or "guy" in America, much as "Mac" or "Bud" (as in "Hey, Bud!" or "Hey, Mac!") has in the twentieth century. So *hillbilly* really just means "hill fellow." This use of "billy" (simply the diminutive of the male name *Bill*) lives on in one other English expression, *billy goat*, which is similarly the generic term for a male goat.

Denizens of Indiana Are Called *Hoosiers.* But What in the World Is a *Hoosier?*

Hoosier, first recorded in 1829, comes from the Cumberland dialect's *hoozer* ("something big"). A probably fanciful theory for the derivation of *hoosier* is that the custom for inhabitants

DAVID FELDMAN

of Indiana, upon hearing a knock on the door, was to inquire, "Who's here?"

Submitted by Mrs. D. L. Billiet of Los Altos, California.

Why Is the Most Violent Sea Called the *Pacific* Ocean?

Those explorers we learned about in elementary school sure seemed to have more guts than brains. Most of them stumbled upon their "discoveries" while looking for somewhere or something else. Perhaps the most famous explorer, Magellan (so famous he goes by only one name, à la Liberace, Cher, and Sting), was responsible for naming El Pacífico ("peaceful," "mild") when he left Spain in 1519 to venture out and explore the New World. As Magellan sailed west, he found the water smooth and the weather pleasant.

Of course, Magellan had no idea he was sailing on a body of water that spans sixty-four million square miles, an area bigger than all of the land on Earth combined. Traveling through the Pacific, one will also encounter the most violent storms of any ocean. But Magellan encountered only the edges of the mighty Pacific. What he didn't see didn't hurt him, so our largest ocean was given a name to indicate its supposed serenity rather than its enormous size.

Why Does Someone Who Swears "Apologize" by Saying *Pardon My French?*

Pardon my French is simply another of the many American and English expressions that equate anything French with sex and obscenity. *French postcards, French novels,* and *French kissing,* just about anything French but French salad dressing, connotes raciness and anti-Puritanism. Come to think of it, you can't find French salad dressing in France, anyway.

Pardon my French started circulating on both sides of the Atlantic around 1916 and so almost certainly stems from the World War I escapades of American and British soldiers.

Submitted by Jean and George Hanamoto of Morgan Hill, California.

DAVID FELDMAN

Why Are Identical Twins with Bodies Congenitally Joined Together Called *Siamese Twins?*

Although Chang and Eng Bunker were far from the first Siamese twins, they were the first to be so named. Born sharing a liver, Chang and Eng were exploited by P. T. Barnum and became stars on the sideshow and carnival circuit of the early- and mid-nineteenth century.

Ironically, although the country of their birth provided the name *Siamese*, they were actually three-quarters Chinese. The Bunkers were able to enjoy a long (1811–74) and relatively happy life. They married twin sisters and fathered twenty-two children between them.

Why Are Many of the Underdeveloped Countries in Asia, Africa, and Latin America Called the Third World? Who are the *First* and *Second Worlds?*

Third World originally referred to nonaligned countries who chose not to be a satellite of either the Communist world (the Second World), or the Western bloc (the First World) on the non-Communist side of the Iron Curtain. This explains why Japan, China, and South Africa, although on the "right" continents, were never called *Third World* countries.

Originally, *Third World* did not connote any negative baggage regarding economic privation, but it gained such an image in the 1960s and 1970s. *Third World* is an apt term because it implicitly acknowledges the pecking order; although Third World countries might provide swing votes in the U.N. General Assembly, they have neither the political nor the economic clout of those countries firmly tied to either of the first two "worlds."

Submitted by Jeff Burger of Phoenix, Arizona.

DAVID FELDMAN

Eponyms: Their Names Are Legion

Who Is the Tetrazzini in *Chicken Tetrazzini?*

This dish, once illustrious, is now less likely to be served at a swanky dinner party than thawed and cooked in a microwave oven. A jumble of diced chicken, spaghetti, cheese, and mushrooms baked in a casserole in a sherry-spiked cream sauce, it was the favorite dish of an Italian diva, Luisa Tetrazzini.

In the early twentieth century, Tetrazzini was world-famous for her portrayal of Lucia di Lammermoor. Luisa was a walking (or, more accurately, a waddling) testimony to the high caloric content of chicken Tetrazzini. The massive soprano evidently didn't stop at Stouffer-size portions of her favorite food.

Who Was the Benedict That *Eggs Benedict* Were Named After?

Rest easy. You aren't unpatriotic if you enjoy this brunch specialty. It wasn't named after Benedict Arnold.

However, *eggs Benedict* were named after a more benign person, a ne'er-do-well member of New York café society who made few, if any, other contributions to our culture.

One morning in 1894, Samuel Benedict staggered into the Waldorf-Astoria and ordered an antidote for his hangover. Charles Earle Funk reports that he ordered "bacon, buttered toast, two poached eggs, and a hooker of hollandaise." The maître d'hôtel, the renowned Oscar, decided to improve on this new dish by substituting ham for bacon, and an English muffin for toast. Oscar honored Benedict by naming the new breakfast after good old Sam.

Why Is a Book of Maps Called an *Atlas?*

Atlas, one of the Titans who tried to overthrow Zeus, was given a rather strange sentence for his offense. He had to spend his life supporting the pillars of heaven on his shoulders. Not being a complete fool, Atlas went to the high mountains of North Africa to do his job, since the peaks were closest to the heavens and he would have less of a burden to bear.

The Flemish geographer Mercator, a pioneer map maker, published his first collection of maps in 1595. Mercator drew a figure of Atlas supporting the world on his shoulders on the title pages. No pillars were to be seen and no explanation for the drawing of Atlas was provided. Although Mercator's opus was not the first published collection of maps, it was the first to be called an *atlas*. Mercator later followed with his most famous collection: *Atlas; or a Geographic Description of the World*.

Who Was the First *Hector* to Bully Someone?

Hector, the Trojan warrior and costar of Homer's *Iliad*. Poor Hector has gotten a bum rap. In the line of duty, he slew Patroclus, a friend of Achilles. This wasn't Hector's smartest move. Achilles slew the brave Trojan in revenge. Hector had never done anything but distinguish himself as a soldier or human. He just happened to have killed the wrong guy, yet he has been saddled, in eternity, with a verb named after him that means "browbeating" and "bullying."

In the early seventeenth century, a London street gang chose to name themselves the *Hectors* after the Trojan hero, who, of course, wasn't around to argue about his name being debased. The Hectors' belligerent behavior is responsible for the verb *hector*.

Why Is a "Take It or Leave It" Proposition Called a *Hobson's Choice?*

Lexicographers love eponyms. When a word or a phrase clearly stems from one fictional or real person, there is none of the ambiguity that marks most etymologies. *Hobson's choice* became a catch phrase when Harold Brighouse wrote a comedy of manners by the same name, but there was also a real Thomas Hobson, who was born in 1544 and who lived to a ripe old age, who clearly was the prototype for this phrase.

Hobson owned a livery stable in Cambridge; he would transport passengers and university mail from Cambridge to London and back. But much of Hobson's business consisted of renting out horses to university students, most of whom were already accomplished horsemen and who had a tendency to overwork the animals. Hobson developed a logical but autocratic system: Regardless of customer preference, a renter could select only the horse nearest the stable door (i.e., the horse who had run the least recently). *Hobson's choice* is often incorrectly used to describe a *dilemma*. A choice of a wide array of wonderful horses would be a dilemma; a *Hobson's choice* implies the appearance of a difficult choice that is, in reality, no choice at all.

Hobson became quite a celebrity in his time. It certainly didn't hurt that so many of his Cambridge customers became powerful figures and literary giants. Hobson continued to work at his livery stable well into his eighties. When he was forced to retire and died soon thereafter, John Milton wrote a humorous epitaph to immortalize him. It read, in part:

> Ease was his chief disease, and to judge right,
> He died for heaviness that his cart went light;
> His leisure told him that his time was come,
> And lack of load made his life burdensome.

DAVID FELDMAN

Why Do We Say That Someone Who "Has It Made" Is *In Like Flynn?* Was There a Real Flynn?

There sure was. Ed Flynn headed the New York City Democratic Party machine in the 1940s. Based in the Bronx, Flynn was a consummate dispenser of patronage. Once you got into his good graces, Flynn could get you elected, get you a cushy job, and maybe even get your trash collected.

Submitted by Leonard Lopate of New York, New York.

WHO PUT THE BUTTER IN BUTTERFLY?   **157**

Why Do We Say That Someone Who Was Rescued at the Last Possible Moment Was Saved *in the Nick of Time?*

In the Middle Ages, wooden tally sticks were used for many purposes that would now be served by paper or by computers. The three most common uses were as ledgers, as records of attendance, and as scoreboards. Until 1826, England used these tallies to record loans to and from the government. The stick was split lengthwise: one side recorded the status of the debtor; the other side, the sum owed the creditor.

In the nick of time probably refers to the practice of schools and churches notching a mark when a student was present. If a young person was running late and just managed to arrive at the classroom as attendance was being taken, he would have been credited, literally nicked (i.e., notched), just in time.

Why Is Joking Around or Fooling Someone Known as *Joshing?*

Unfortunately, not even eponyms can always be clearly traced to one source. *Josh* is one such example.

At first blush, the derivation of *Josh* seems clear. An extremely popular humorist of the nineteenth century, Henry Wheeler Shaw, wrote newspaper columns under the pseudonym Josh Billings. His columns, starting in the 1860s, were full of wordplay—puns, malapropisms, misspelled words, and intentional illiteracy stated in a generic southern dialect. He was to Mark Twain what the Ritz Brothers were to the Marx Brothers. Without a doubt, Shaw's kidding around in his col-

DAVID FELDMAN

umns helped spread the current meaning of *josh*. But as hard as we try to attribute the coining of this expression to Shaw, it doesn't fit. For *josh* had been cited in print, with the same meaning, in 1845. So where does *josh* come from?

Our usual "reliable source," the *OED*, is notably mum on the subject. But theories abound:

1. *Josh* is a contraction of *joke* and *bush*.
2. *Josh* is descended from the Scottish word *joss*, meaning to jostle or push around.
3. *Josh* was slang for a man from Arkansas (this expression gained favor during the Civil War, so it is unlikely to be the genesis of the current meaning).
4. *Josh* is related to the English *joskin*, meaning bumpkin, a good description of the kind of characters Josh Billings wrote about.

Why Is Straining to Live in the Style of One's Neighbors Known as *Keeping Up with the Joneses?*

This phrase was not named after a real *Jones*, but it was created by someone with an unpleasant experience of trying to keep up with the neighbors. Arthur R. Momand, an artist earning an impressive six thousand dollars a year in the early twentieth century, left New York City for the tonier climes of Cedarhurst, Long Island. He soon found that his neighbors' average income was double his, and his attempt to match their standard of living was leading to frustration and drastic depletion of his bank account. Momand and his wife moved back to the Big Apple.

Any artist worth his salt is willing to make a buck out of adversity. Momand saw that his attempts to accommodate to Robin Leach life-styles on a middle-class income was rather silly and great fodder for a comic strip. Momand decided to draw a comic strip, *Keeping up with the Smiths*, based on the theme of futile social overreaching. At the last minute, Momand decided *Joneses* sounded better than *Smiths*. The strip was an instant success and spawned a musical comedy, two-reelers, and book anthologies.

In England, *keeping up with the Joneses* didn't become a catch phrase until well after World War II. The phrase caught on when a "commoner," Anthony Armstrong-*Jones*, had the "audacity" to marry Princess Margaret. In England, then, *keeping up with the Joneses* implies social climbing, while in United States it refers more to trying to mimic the outer trappings of affluence, a rather accurate barometer of the importance of class and materialism, respectively, in the two countries.

Who Was the Newburg That *Lobster Newburg* Is Named After?

Lobster Newburg wasn't named after a Newburg. This rich dish, made with a sauce of sherry, heavy cream, and egg yolks, was originally named *lobster Wenberg*. In the 1890s, customer Ben Wenberg, a shipping magnate, showed the chef at Delmonico's (New York's most elegant restaurant) how to prepare a similar South American dish he had sampled.

Like the Benedict who lent his name to *eggs Benedict*, Wenberg was far from an angel. In fact, Delmonico's was forced to expel Wenberg after a particularly ugly drunken brawl and decided not only to banish the patron, but his name from the menu as well. The first syllable of Wenberg's name was transposed and *Lobster Wenberg* was rechristened *Lobster Newburg*.

DAVID FELDMAN

What in the Heck Is a *Love Jones?*

I'll never forget the first time I heard the song by Brighter Side of Darkness, *Love Jones*. A young, male voice with a startling falsetto sang a standard plaintive ballad of longing and loss. But the insistent refrain "I've got a love jones for you" was the hook. Nobody who heard the song could fail to understand the meaning of the song: This poor boy *really* loved this girl. But what the heck is a love jones?

Cheech and Chong made fun of the song with their wonderful *Basketball Jones*, yet it is the innocence, earnestness, and intensity of the original that sticks in people's minds. When I've done radio shows promoting *Imponderables*, I've frequently been asked, "What is a love jones?"

The answer is simple. A *jones* is a habit, usually a reference to a serious drug habit, and although the term started among heroin users, it spread to black slang. The fervor with which the boy in *Love Jones* longs for his girl approaches that of a junkie for his needle. Nobody seems to know who the *Jones* was who was given the dubious honor of lending his name to this expression.

Submitted by N. T. B. of Oakland, California.

Which Tom Was the First *Peeping Tom?*

Leofric, earl of Mercia and lord of Coventry, a nobleman in the eleventh century, acted more like a 1950s situation comedy husband than an ancient autocrat. He imposed a heavy tax on his citizens, and his wife protested. "O.K.," he said, "I'll abolish the tax under one condition: Ride around the town nude on a horse."

"No sweat," said his better half, who was Lady Godiva, by the way. So, according to legend, Godiva ordered the citizens to stay inside their homes and cover their windows. Everyone obliged except for one Coventry tailor named Tom, who couldn't resist a little peek.

Poor Tom. His punishment was severe, much worse than a tax. He was struck blind for his little indiscretion and has lived forever as the symbol of voyeurism.

DAVID FELDMAN

Which *Nick* Lent His *Name* to *Nickname?*

No, it was not the Nick from "just in the nick of time." *Nick-name* goes back to the Middle English (1303) *ekename* ("additional name"). Evidently, *nickname* was formed by slurring the *n* in *an*. If you notice, saying *"an ekename"* is awkward indeed.

Ekename also was the source for *eke* (not what you yell when you encounter a mouse, but the word that means "to supplement"—especially to supplement income).

Submitted by Pete Eisenhauer of Fort Lauderdale, Florida.

Who Is the *Peter* Responsible for Failing or Tapering Off in the Expression *Peter Out?*

If there was one real person responsible for this aspersion on all Peters, it was the Apostle Peter. When Jesus was arrested in the Garden of Gethsemane, Peter, at first, attempted to reclaim Jesus by force, but eventually not only gave up attempts to rescue his teacher but also denied "three times before the cock crowed" that he even knew Jesus. *Petering out* usually implies this kind of deterioration of effort, but the expression did not become popularly used until the nineteenth century, raising great skepticism about the apostle's failures as the source.

We do know for sure that *peter* was a popular term in North American mining camps in the mid-nineteenth century. There were two techniques to obtain gold: superficial panning, called "placer mining"; and deep mining, involving explosives. These explosives contained saltpeter. After a seam of a mine was depleted of all possible gold, it was referred to as "petered out." The miners then went somewhere else and exploded a new seam.

We would be derelict in our responsibilities if we didn't add a third theory for the genesis of *peter out,* and it is more plausible than the first: Our word might be a direct descendant of the French word *peter,* which means "to fart." One could hardly ask for a word that better connotes "fizzling out."

Was There Ever a Problem with un-*Real McCoys?*

One of the hallmarks of a folk etymology is remarkably similar stories with similar payoffs occurring in geographically disparate parts of the country (or world) at about the same time. We may never know for sure where *the real McCoy* comes from, but on one thing all agree: This cliché has nothing to do with the feuding McCoys of "Hatfields and McCoys" fame.

The real McCoy was first recorded at about the turn of the twentieth century. Most etymologists agree that the McCoy referred to was an eminent boxer of the era, Norman Selby, whose ring name was Kid McCoy. McCoy was a phenom in the 1890s, quickly becoming the world's welterweight champion. One story, which reeks with improbability, posits that McCoy engaged in a barroom brawl with a boisterous patron who refused to believe that Selby, no giant, could really be the famed ring terror. As he picked himself off the floor, the victim realized that Selby must be *the real McCoy.*

Far more likely is that Selby was beset by McCoy imitators. Much as occasional bogus groups masquerade as genuine 1950s doo-wop acts in oldies shows, barnstorming pugilists in Selby's time used to "adopt" the names of famous boxers to

DAVID FELDMAN

entice larger box office. Elvis Presley might not begrudge the slew of his imitators, but none of the clones bills himself as the King. There were so many boxers named Kid McCoy on the barnstorming circuit that, so the story goes, Selby had to rename himself "Kid 'The Real' McCoy" to differentiate himself from his simulators.

Among others, John Ciardi, has found the Kid McCoy explanation too pat, as most folk etymologies are. Ciardi thinks it is more likely that this cliché is actually a reference to an excellent Scotch whisky, Mackay, which, to confuse things further, was imported into the United States just slightly before Kid McCoy hit his stride as a boxer. Mackay was a brand preferred by Scotsmen in America, and it carried a certain nineteenth-century "designer label" cachet. With typical ingenuity, entrepreneurs marketed mislabeled swill as the genuine article. *The real Mackay*, then, was a reference to the right stuff—genuine, imported Mackay Scotch.

Possibly, a phrase originally created to describe the whisky was applied to the boxer.

Who Was the Drink *Tom Collins* Named After?

The exact origins of the *Tom Collins* is obscure, but Tom Collins, a bartender at Limmer's Old House in London, specialized in mixing the meanest combination of gin, lemon or lime juice, sugar, and soda water on ice. Some credit the bartender for creating the drink that bears his name, but this honor is widely disputed.

Why Is a Lightweight Automatic Machine Gun Called a *Tommy Gun*?

Tommy guns were named after a John, John T. Thompson, the head of the Small Arms Division of the U. S. Army during World War I. Thompson and Navy commander John Bish worked on prototypes during the war and made many modifications after the war.

Tommy gun eventually became a generic term for any lightweight automatic machine gun with a drum-type magazine. Although we associate tommy guns with mobsters on *The Untouchables*, they were also used by Allied troops during World War II.

Was the London *Bobbie* Named After a Real Bobby?

Yes—Bobby, or rather, Sir Robert, Peel, who established the London Metropolitan Police in 1829.

Who Was the Robert That *Bobby Pins* Were Named After?

Bobby pins were not named after a person, but a haircut—the short bob hairstyle popularized during the Roaring Twenties by flappers.

Submitted by Mrs. Harold Feinstein of Skokie, Illinois.

WHO PUT THE BUTTER IN BUTTERFLY?   **167**

Food Words for Thought

Has Anyone Ever Literally *Eaten Humble Pie?*

Yes—and voluntarily, too. In the sixteenth century, huntsmen ate more than their pride when they consumed *humble pie.* *Humble pie* was originally *umble pie,* a pastry made out of the heart, liver, and entrails of wild animals, usually deer. After a deer hunt, while the nobility enjoyed filets of venison in the master dining rooms of their palaces, the huntsmen had to content themselves with the more "umble" offering.

By the nineteenth century, *umble* became *humble.* James Rogers speculates that the transformation was self-conscious wordplay "on the humble station of people who ate umble pie."

Why Is a Bad Actor, Particularly One Who Overacts, Called a *Ham*?

We figured that *ham* must have been created by a disillusioned director. Alfred Hitchcock once remarked that actors should be treated like cattle. Could pigs be far behind?

Actually, *ham* is an abbreviation of *hamfatter*, a term used to describe second-rate performers who were prone to exaggerated gestures in minstrel shows during the mid-nineteenth century. Minstrels blackened their faces with burnt

DAVID FELDMAN

cork and removed their makeup with ham fat rather than the cold cream that more affluent actors could afford.

Another theory speculates that *ham* is attributable to Shakespeare's Hamlet, who railed against the type of acting we now call "hammy." The strongest argument against the *Hamlet* theory is that the expression was not popularized until the 1880s, a time when minstrel shows, which originated in 1842, were still popular.

A few lexicographers have suggested that *ham* is a variant of *amateur*. This is probably how the expression *ham radio operator* originated. Not only is there no evidence of this adaptation, but also the nuances of how *ham* is used are missed. A *ham* isn't necessarily technically deficient, and we have plenty of words to describe ineptitude. But a word was needed to describe the Bert Parkses and Sammy Davis, Jr.'s, of the world, entertainers capable of quality performances and incapable of withholding nods and winks.

Why Is the Cereal Called *Grape-Nuts* When It Contains Neither Grapes nor Nuts?

C. W. Post introduced this cereal in 1898 and dubbed it *Grape-Nuts* because of the natural sweetness of wheat and malted barley ("sweet as grapes," he said) and because it was as crunchy as nuts.

Crunchy as nuts? The cereal is much crunchier than nuts unless you soak it in milk for a few months. Wouldn't *Grape-Rocks* have been more appropriate?

Why Would Anyone Who Has Ever Baked Say Something Was *Easy as Pie?*

Anyone who has ever contended with making a pie crust from scratch will find this metaphor particularly inapt. But the origin of this cliché explains the conundrum. *Easy as pie* is a contraction of a late-nineteenth-century catch phrase, *easy as EATING pie.* Now *that* makes some sense.

Why Do We Call That Thing in the Backyard a *Barbecue?*

The first recorded use of *barbecue* was in 1661, clearly a borrowing from the Spanish *barbacoa* (meaning "framework of sticks"). Originally, a barbecue was the raised device used to roast whole animals over open fires. Now we use *barbecue* to refer to the cooking device, the meat so cooked, and even the sauce poured over it.

Why Do We Call Liquor *Booze?*

Both England and the United States can lay some claim to this expression. The Middle English *bousen* meant "to drink deeply" or "carouse." The English used *bouse* to refer only to beer or ale, which is curious, since in the United States, *booze* usually refers to "hard" liquor.

So we are left with the amazing coincidence that there was a Kentucky colonel distiller named Booze who sold whisky under his own name. Booze marketed his booze in glass containers in the shape of log cabins. In 1840, the colonel was blessed with a stroke of luck. General William Henry Harrison ran against Martin Van Buren by lashing out against Van Buren's blue-blood roots. Harrison's harping on his humble origins ("I was born in a log cabin") not only helped elect Harrison but also proved a boon to Booze's business.

Where Does the *Cole* in *Coleslaw* Come from?

Illiterate menus to the contrary, *cole* has nothing to do with *cold*. *Coli* means "cabbage" in Latin and *sla* means "salad" in Dutch. The Dutch word *koolsla*, appropriately enough, means cabbage salad.

Submitted by Mrs. Harold Feinstein of Skokie, Illinois.

Why Does *Cut the Mustard* Mean "to Succeed or Meet Expectations"?

Our current *cut the mustard* is clearly the descendant of *to be the proper mustard* ("O.K.," "genuine") and *all the mustard* ("great"), two early-twentieth-century expressions. The first recorded use of *cut the mustard,* in 1907, is from a O. Henry short story: "I looked around and found a proposition that exactly cut the mustard."

Where did the earlier expressions come from? Most likely, *mustard* is a corruption of *muster.* Anyone who can *cut the mustard* can clearly *pass muster.*

Another intriguing theory is that the *cut* in *cut the mustard* is used to mean "dilute" (as in "the street heroin was *cut* with flour"). *Cut the mustard* could refer to the necessity of diluting mustard powder with vinegar to provide relief for the palate.

No doubt, the spread of prepared mustard in the early twentieth century hastened the popularity of all the "mustard phrases."

Why Do We Call a Glass of Liquor a *Highball?*

Most likely, *highball* derived from late-nineteenth-century bartenders' lingo that called glasses *balls. Highball* replaced another expression, *long drink,* which at first referred specifically to Scotch and soda and later applied to any whisky and soda served in a tall glass.

For the record, Charles Earle Funk believed *highball* was a signal to the locomotive engineer that it was safe to bypass a station without stopping. The signal was a large ball that would be hoisted to the top of a mast to indicate that there were no passengers, freight, or connecting trains awaiting it. Funk had a harder time explaining how the expression switched from the railroads to the barroom. Funk's best speculation was that a drunk train passenger might have noticed the resemblance between the floating ice atop a high glass and the ball atop the train signal. Not likely.

Why Is Someone Who Works Behind a Fountain Called a Soda *Jerk?*

Around 1800, in America, drunkards were called *jerkers,* presumably because of their unstable gait. Beer fanciers were dubbed *beer jerkers.* By at least 1873, *beer jerker* was the title bestowed upon dispensers of suds. With the spread of soft drinks in the early twentieth century, fountain employees were honored with the appellation as well.

DAVID FELDMAN

Is It *Ketchup* or *Catsup?*

It's *ketchup,* even though the English sailors who brought the condiment back from Singapore in the seventeenth century didn't have the slightest idea how the word should be spelled.

The original ketchup was the Chinese *ke-tsiap,* a pickled fish sauce. The Malays stole the name *(kechup)* but not the base—they used mushrooms instead of fish.

Americans added the tomatoes, and Heinz's first major product, tomato ketchup, was launched in 1876. Since the Chinese, Malay, English, and American incarnations all began with the "ke" sound, most word purists would rather say "hopefully" indiscriminately than be caught dead spelling the word *c-a-t-s-u-p.*

WHO PUT THE BUTTER IN BUTTERFLY?

Why Is a Mess or Confusion Called a *Pretty Kettle of Fish?*

The Scots knew how to throw a picnic. No hot dogs or fried chicken for them. In the eighteenth century, the picnic season started at the beginning of the salmon run each year, and picnics were conducted along the banks of the river. Salmon was plentiful, but they didn't have an elegant solution to the problem of how to prepare the fish. They cooked their catches in large kettles and tried to eat their hot boiled salmon with their fingers. It was a mess.

The *pretty* preceding *kettle of fish,* then, always has been ironic. An American phrase, *pretty picnic*, has exactly the same meaning.

Why Is Youth, Especially the Golden Days of Youth, Called the *Salad Days?*

No, *salad days* are not a sale at the local salad bar. *Salad days* can be a synonym for "youth," but the connotation is not only greenness and inexperience but also a peak time of life that can never be recaptured (think *Summer of '42*).

William Shakespeare coined this expression, and even explained the metaphor, in *Antony and Cleopatra*. Cleopatra says, in Act 1, Scene 5:

> My salad days
> when I was green in judgement: cold in blood,
> To say as I said then!

Why Would Anyone Want to Call an Edible Substance *Shoofly Pie?*

Because they are trying to shoo away flies, who are inordinately attracted to the stuff, from messing with their open pie made of a sugar and molasses filling.

Submitted by Maurice H. Williams of Stewartstown, Pennsylvania.

Why Do We Call Children *Small Fry?*

A caller on a talk show asked why we use the same expression to describe our children and a side order at McDonald's. *Fry,* or variants of it, has for centuries meant "children," starting with the Old Norse *frjo* ("children of a man's family") and continuing in Middle English usage. What confuses the issue is that in American slang, we mix up *fry* with *potato,* and *small potatoes,* a pejorative term, has been around since 1831.

Small fish are still commonly referred to as *fry,* and most dictionaries include "children" as one of the current definitions of *fry.* Interestingly, *small fry* is used affectionately when describing offspring, but usually disparagingly when applied to things. ("He talks like he is running IBM, but it is just a small-fry operation.")

Where Did the *Lollipop* Get Its Name?

From the British. *Lolly* is English dialect for "tongue." The *pop* undoubtedly comes from the lip-smacking noise of the sucker (or should we say, the *licker,* since *sucker* can refer to the "licker" or the "lickee").

In England, *lolly* has long been slang for candy. In North America, the *lollipop*'s place in the candy pantheon was cemented in the 1930s when Shirley Temple warbled "On the Good Ship *Lollipop*."

Why Is an Ice Cream with Syrup Called a *Sundae?*

Two states vie for the honor of claiming *sundae*. We will probably never settle the issue. H. L. Mencken bestowed the honor upon Wisconsin in 1890. George Giffy owned an ice cream parlor in Manitowoc. Giffy's best seller was plain vanilla ice cream, but a customer tipped off Giffy to the wonders of pouring chocolate syrup, then used primarily as a flavoring for sodas, on ice cream. Giffy saw the potential in the creation but felt that he had to justify the five cents extra he would charge for his new confection. Giffy's solution was to sell his ice cream and syrup on Sunday only, when post-church business filled the place. Possibly to avoid blasphemy by associating his dessert with the Sabbath, or just the combination of pretension and contrarianism that drives us to call *nightclubs*, *niteclubs*, Giffy changed the spelling to *sundae*.

Norfolk, Virginia, also enters a claim in the sundae sweepstakes. Norfolk had extremely restrictive blue laws that prohibited the sale of not only alcohol but of soda drinks as well. A clever fountain owner concocted the ice cream and syrup combination to circumvent the blue laws. By creating a "dry" soda (what is a sundae, after all, but an ice cream soda without the soda?) and naming it after its birthday, Sunday, he was able to stay open and peddle his wares on the Sabbath.

Why Is a Speech Delivered Before Drinking Called a Toast?

In the seventeenth century, long before wine coolers or Harvey Wallbangers, revelers put pieces of spiced, toasted bread in wine (or, less frequently, in ale or sherry). Soon it became *de rigueur* to drain a glassful when someone was saluted. One who drank all the wine thus consumed the toast as well.

Neil Ewart reports an almost certainly apocryphal anecdote to explain the origin of *toast:* A gentleman eyed a beautiful woman at the public baths. Inspired, he dipped a glass into the water, held it up to public view, and drank to her health. "Whereupon a reveller, who had been enjoying more real liquor than the natural mineral water, jumped into the bath and declared that he would have nothing to do with the liquor but would have the toast . . . in other words, the lady herself."

DAVID FELDMAN

Why Is Indecision Called *Waffling?*

Despite the attempt by some etymologists to equate the confusion and equivocation of *wafflers* with the complex, honeycombed patterns of waffles, the evidence lies elsewhere. In Old English, the word *wafian* means "to wave."

As early as 1560, *whiffle* was recorded in England to describe shifts of wind direction (e.g., "to *whiffle* like a weather vane"). This meaning undoubtedly inspired the whiffle balls of the 1950s, whose trajectories after they were thrown were less predictable than a weather vane's. In the 1960s, *whiffle* was transformed into *waffle,* and the perfect term to describe recalcitrant Senate subcommittee witnesses was born.

Local and Other Colors

Why Is a Diversion or Distraction Called a "*Red Herring*"?

The cliché "neither fish nor fowl" is actually shortened from another expression, "neither fish nor fowl nor good red herring." William and Mary Morris attribute the earlier expression to the dietary caste system of the Middle Ages. It was then believed that only the clergy were worthy of eating fish. The masses should be happy with fowl. And paupers would have to be happy with red herring. (Royalty, of course, could eat whatever they damn pleased.) *Neither fish nor fowl nor good red herring,* then, referred to something that wasn't suitable for anybody.

But why a *red* herring? Refrigeration was nonexistent in the Middle Ages, so paupers sun-dried and salted herring, and the fish turned dark. But those who also smoke-cured the herring found that the fish turned a bright red color. Hence, the red herring.

Smoked herring was popular among a few other groups as well. Sailors found smoked herring the perfect food to carry on long voyages (much as American cowboys relied on beef jerky), for it would remain edible long after fresh meat or fish.

But another, more elite group also found a valuable use for red herring: hunters. Smoked herring has a strong odor, and hunters found that it was the perfect substance to train young bloodhounds. Before dogs were expected to follow the tracks of foxes, hunters would drag the herring along a trail. If the hound showed talent at discerning the "herring trail," it would advance to real chases.

Two other groups, with more complicated motives, also used red herring to promote their purposes. Criminal fugitives in the seventeenth century would drag red herring to divert the trail of bloodhounds in pursuit. And animal-rights groups would sabotage hunting expeditions by laying red herring along the path of the fox-chasers.

Fooling the fox became known as "faulting the hounds." If a hound was diverted from a trail by a false clue, it would follow a real red herring. When we are distracted by a deliberately laid trap, we follow a metaphorical red herring.

Why Is a Lucky or Special Day Called a *Red-Letter Day?*

As early as the fifteenth century, ecclesiastical calendars designated religious holidays by printing them in red (and, less frequently, purple) letters. In England, saints' days and feast days were printed in red in the calendar of the Book of Common Prayer, indicating that special services were provided for these days. Many calendars distributed by churches and civic organizations still print holidays, and often Sundays, in red.

Why Is Excessive Bureaucratic Formality and Delay Called *Red Tape?*

Metaphorical red tape seems to exist in any country that has a bureaucracy—that is, everywhere. But actual red tape was once the tangible symbol of a government's exasperating tendency to prolong the simplest transaction.

English lawyers and government officials had traditionally tied official papers together with red ribbon, which they called *red tape* even though it didn't contain an adhesive. Papers were delivered rolled up with the distinctive red ribbon announcing the importance of the documents. What exasperated Charles Dickens and Thomas Carlyle, who popularized this expression, was that these papers were again tied up with ribbon after every use, even when they were shelved for storage. Retrieving any official papers required the elaborate procedure of untying and eventually retying the red tape, a small but irritating and time-consuming inconvenience.

Why Is Money Paid to Prevent a Secret from Being Exposed Called *Blackmail?*

Blackmail is a Scottish expression that dates back to the sixteenth century. Northern areas of Scotland were subject to the plundering of pirates and other freebooters. Highland chiefs used to exact a tribute from landholders in exchange for protection against the looting of freebooters.

The word *mail* had long meant "rent" or "tribute." Even today, a Scottish tenant is called a "mailer." But why *blackmail?* Two different forms of "mail" were made by tenants: *Whitmal* (i.e, whitemail) was payment in silver; *blackmal* (blackmail) was payment in labor, cattle, grains, or produce. As silver was considered a more desirable commodity than sacks of corn, *blackmail* seems to be another example of the use of *black* to denote "lower" or "inferior."

Many renters still feel that their monthly payments are more tribute to the landlord than fair compensation for worthy value, but today the word *blackmail*, has different connotations. The original Scottish term describes what we would now call a protection racket. The connotation of blackmail as the threat of exposure of a secret, and the payment as a bribe or hush money, was not widespread until the nineteenth century.

Why Do We Turn *Green* with Envy?

Judith S. Neaman and Carole G. Silver report that *green* and *pale* were alternate meanings of the same Greek word. In the seventh century B.C., the poetess Sappho, used the word *green* to describe the complexion of a stricken lover. The Greeks

DAVID FELDMAN

believed that jealousy was accompanied by an overproduction of bile, lending a pallid green cast to the victim.

Ovid, Chaucer, and Shakespeare followed suit, freely using *green* to denote jealousy or envy. Perhaps the most famous such reference is Iago's speech in Act 3 of *Othello:*

> O! beware my lord, of Jealousy;
> It is the green-ey'd monster which doth mock
> The meat it feeds on.

Although we are now more likely to ascribe the pallor of a friend to a questionable tuna fish salad sandwich rather than an emotional fit, *green with envy* remains entrenched.

Submitted by Tony Drawdy of Bamberg, South Carolina.

Why Is an Inexperienced Person Called a *Greenhorn*?

Green has long been associated with things young and immature. The question is: Why "green*horn*"?

John Ciardi supplies the only plausible theory we have encountered. When young deer grow horns, a temporary skin surrounds the horn to protect the delicate growth (much as husks grow around the shells of nuts). Greenish fungus spores tend to grow on this temporary skin, which make the horns look green until the skin peels away completely.

• Head of a young girl with pinkje •

Why Is Our Little Finger Called a *Pinkie?* Why Is the Drink Called a *Pink Lady?*

We have the Dutch to blame for this piece of baby talk. *Pinck* in Middle Dutch meant "small." Their word for the little finger was *pinkje.*

At least a *pink lady* is pink (although why it is called a "lady" probably has more to do with sexism—froth equals femininity—than linguistics). John Ciardi points out that the diminutive ending of *pinkie* is a redundancy, for it means, in effect, "small little finger." At least the Dutch derivation explains why one finger has been singled out for its pinkness.

Submitted by Steve Hajewski of New Berlin, Wisconsin.

Why Do We Say, "Not a *Red* Cent"?

Remember the days of yore (whatever the heck *yore* is) when U. S. coinage was made of real, identifiable metals? Pennies used to be made out of copper but now are an alloy of copper, tin, and zinc. The "red" in "red cent" was a reference to the hue of the original copper penny; the current alloy contains reddish but also golden and brown hues.

Why Is an Unwanted or Useless Possession Called a *White* Elephant?

The quintessential *white elephant* has several qualities in common: It is large and unwieldy; it is expensive (or at least extravagant for its category); it is a gift; and it is a gift that can't be refused in the first place and then can't be returned or destroyed for some reason. (Otherwise, who would want to keep a white elephant around the house?)

All of these connotations stem from the original *white elephants,* a strain of albino (actually whitish-gray) pachyderms that were considered sacred by the Siamese. Any captured white elephant, became, by law, property of the emperor. Under no circumstances could white elephants be destroyed without royal permission.

Any monarch worth his salt is besieged by assorted hangers-on, and a Siamese emperor devised an ingenious method to punish particularly obnoxious courtiers. He bestowed upon them the gift of one of the sacred white elephants.

Unsuspecting courtiers probably could think only about the good news: the tremendous honor they had received. Rather soon, however, they recognized the bad news: They were saddled with a literal and metaphorical *white elephant.* At least that awful vase that your boss brought you when you invited him over for dinner is an inanimate object, one that can be dragged out of the basement or a closet if he should by some miracle ever be reinvited back. But real elephants do annoying things like eating and defecating and running around and tearing down fences.

Recipients of the white elephants were not allowed to work the elephants. They couldn't even ride them. Only the emperor was allowed to ride the white elephant, and the recipient was always aware that he must keep the elephant on hand in case the emperor decided he felt like a trot.

But the recipients had a more pressing problem: They simply could not afford the upkeep of the elephant. So although the contemporary recipient of a white elephant is only emotionally scarred, the earliest victims were inevitably financially ruined.

DAVID FELDMAN

Oddballs: Words Whose Only Deficiency Is Their Inability to Fit into Any of the Other Chapters

Why Is Every Fourth Year Called a *Leap Year?*

This eighth-century English idiom has always seemed inappropriate. The year itself doesn't leap; an extra day is tacked on to the year. "Leap-day year" might have been a more appropriate name.

But there is a logical reason for the name. Divide the 365-day typical year by the 52 weeks. You will see that there is one additional day. Thus, in nonleap years, if a fixed-date holiday, such as Christmas, was held on a Tuesday, one knew that the next year it would fall on a Wednesday. But in a leap year, the festival would fall two days after the previous year's. This skipping of a day is the "leap" in *leap year.*

The year 2000 will be an exciting one for leap-year enthusiasts, for centennial years are leap years *only* when divisible by 400. The last centennial leap year was 1600, and it is doubtful that most of us will be around to enjoy the next one after 2000, in 2400.

Why Is a Two-Week Period Called a *Fortnight?*

The Anglo-Saxons believed that the night was a discrete entity from the daytime, and they measured the passage of time by counting the number of nights that had passed. *Fortnight* comes directly from the Old English *feowertene nighta* ("fourteen nights").

Fortnight has a quaint and obsolete feel to it, but no other English word is more precise. Those of us who subscribe to many magazines and journals are plagued by descriptions of their frequency: Is a *bimonthly* magazine one that is published twice a month, or once every two months? Is there necessarily a difference between *bimonthly* and *biweekly?*

In the best of all possible worlds, *biweekly* would mean "once every two weeks." We have another term, *semiweekly*, which is perfectly adequate to describe something produced twice a week.

Even if we could standardize terms, a further problem still exists: There are more than four weeks in every month except for February (in nonleap years). A biweekly magazine would come out twenty-six times a year, but a semimonthly only twenty-four times a year. Subscribers aren't always sure how many issues they are receiving.

The simplest solution, I propose, is to bring back *fortnightly* as the term to describe "every two weeks." The word has been around for about fifteen hundred years, longer certainly than any of these magazines confusing us about their frequency.

Submitted by Pam Lebo of Glen Burnie, Maryland.

Do *Gunnysacks* Have Anything to Do with Weapons?

Nothing whatsoever. *Gunnysack,* imported into the United States around the Civil War, is an Anglicization of the eighteenth-century Hindu-Sanskrit *goni.*

What we now call *gunnysacks* were originally made in Bengal. At first, *goni* referred to the material (jute and hemp) that was used to make the sacks, and then to the sacks themselves.

What did *goni* mean in Hindu? "Sack." That's right. Americans have been running "sacksack" races at picnics for over a century now.

Why Is Groping Around to Find the Right Words Called *to Hem and Haw?*

Look up the word *hem* in your dictionary. The first definition will be something like "the border of a garment" or "a margin." A subsequent definition will be something like "noun, the sound of clearing the throat."

Haw has several different meanings. (Did you know that a *haw* is a berry—the berry of a hawthorn tree?) One meaning, according to *Webster's New World Dictionary,* is "a conventionalized expression of the sound often made by a speaker when hesitating briefly."

So we evidently arrived at *hem and haw* by combining the actual sounds of two ways we stall when we don't know what to say or are nervous about saying it. Both are examples of *onomatopoeia,* words formed by imitating the sound associated with the action or object being named. *Buzz,* for example, is simply the attempt to combine letters to echo the sound of a bee. *Tinkle,* with a Middle English lineage, was an attempt to sound like a small bell.

Onomatopoeia, a popular word in spelling bees, has Greek origins and clearly doesn't sound like *any* thing or person. The Greek *onomatopoiia* meant "to make words or names."

Is a *Dandelion* Named After Lions?

Yes. The English had long called the dandelion a *lion's tooth,* but in the sixteenth century, for some reason, they adopted the French name for the flower, *dent de lion,* which, appropriately enough, means "lion's tooth." Why the English would choose to borrow a literal French translation of a perfectly fine English expression and then proceed to mangle both the spelling and pronunciation of *dent de lion* I leave to an Anglophile prepared to explain the glories of English cuisine and the popularity of Rick Astley.

Why would France and England call this flower, which is yellow but looks nothing like a lion's tooth, a *dandelion?* The name refers not to the flower itself but to the surrounding leaves which are deeply indented and resemble teeth.

DAVID FELDMAN

What's Right? *Cattycorner? Kittycorner?* or *Catercorner?*

Catercorner has the pedigree. All are American regional variants of the French *quatre* ("four"). *Catercorner* was first recorded in 1519; its meaning, "the point diagonally across a square or intersection," is exactly the same today as it was then.

Catercorner has nothing whatsoever to do with cats, kitties, or any other felines, and the two alternatives represent only a few of the many regional variants (my favorite of which is the South's *caterwampus*).

Why Do We Say a Sudden Onslaught Was Executed in *One Fell Swoop?*

Fell and *swoop* are both Middle English words, and the two words were combined to describe how a bird dives to capture its prey. The graceful and skillful action of a bird seems to contradict the meaning of *fell*, but "fell" has more than one meaning. The *fell* in *one fell swoop* is derived not from the verb meaning "to fall," but from the same Middle English word, *fell*, that formed the root of *felon.*

Fell in Middle English meant "cruel" or "terrible." *Swoop* meant "snatch." *One fell swoop*, then, accurately described the ruthless proficiency of birds of prey.

Why Is a Pauper's Burial Ground Called a *Potter's Field?*

After his betrayal of Jesus, Judas was left with thirty pieces of silver. *Matthew* 27:7 recounts that Judas, in remorse, flung the precious silver down to the ground. As any deliberative body is wont to do, the elders ruminated long and hard about how to spend the thirty pieces of silver that Judas had provided. "And they took counsel, and bought with them [Judas's thirty pieces of silver to] the potter's field, to bury strangers in. Wherefore the field was called the field of blood. . . ." And Judas's thirty pieces of silver were forever more referred to as *blood money.*

Although the Bible doesn't specify so, *potter's field* probably got its name from the fact that before it was turned into a cemetery, potters had obtained their clay from this field. The original potter's field lies outside Jerusalem and still is called *Aceldama* ("Field of Blood").

Bibliography

Boycott, Rosie. *Batty, Bloomers and Boycott.* New York: Peter Bedrick, 1983.

Brandreth, Gyles. *The Joy of Lex.* New York: Quill, 1983.

Brewer's Dictionary of Phrase and Fable. New York: Harper & Row, 1970.

Chapman, Robert. *New Dictionary of American Slang.* New York: Harper & Row, 1986.

Ciardi, John. *A Browser's Dictionary.* New York: Harper & Row, 1980.

————. *Good Words to You.* New York: Harper & Row, 1987.

Dohan, Mary Helen. *Our Own Words.* New York: Alfred A. Knopf, 1974.

Espy, Willard R. *O Thou Improper, Thou Uncommon Noun.* New York: Clarkson N. Potter, 1978.

Ewart, Neil. *Everyday Phrases: Their Origins and Meanings.* New York: Sterling Publishing, 1983.

Fernald, James C. *Synonyms, Antonyms and Prepositions.* New York: Funk & Wagnalls, 1947.

Flexner, Stuart Berg. *Listening to America.* New York: Simon & Schuster, 1982.

Follett, Wilson. *Modern American Usage.* New York: Hill & Wang, 1966.

Freeman, Morton S. *A Treasury for Word Lovers*. Philadelphia: ISI Press, 1983.

Funk, Charles Earle. *Heavens to Betsy! & Other Curious Sayings*. New York: Perennial Library, 1983.

———. *A Hog on Ice & Other Curious Expressions*. New York: Colophon, 1948.

———. *Thereby Hangs a Tale: Stories of Curious Word Origins*. New York: Perennial Library, 1950.

———and Charles Earle Funk, Jr. *Horsefeathers and Other Curious Words*. New York: Perennial Library, 1958.

Funk, Wilfred. *Word Origins and Their Romantic Stories*. New York: Bell Publishing, 1978.

Garrison, Webb. *What's in a Word?* Nashville, Tenn.: Abingdon Press, 1965.

Hayakawa, S. I. *Choose the Right Word*. New York: Perennial Library, 1987.

Hendrickson, Robert. *The Dictionary of Eponyms*. New York: Stein and Day, 1972.

———. *Salty Words*. New York: Hearst Marine Books, 1984.

Holt, Alfred H. *Phrase and Word Origins*. New York: Dover, 1961.

Hunt, Cecil. *Word Origins: The Romance of Language*. New York: Philosophical Library, 1962.

Laird, Charlton. *The Word*. New York: Simon & Schuster, 1981.

Mencken, H. L. *The American Language*. New York: Alfred A. Knopf, 1937.

Moore, John. *You English Words*. Philadelphia: J. B. Lippincott, 1962.

Morris, William, and Mary Morris. *Harper Dictionary of Contemporary Usage*. New York: Harper & Row, 1985.

———. *Morris Dictionary of Word and Phrase Origins*. New York: Harper & Row, 1977.

Neaman, Judith S., and Carole G. Silver. *Kind Words: A Thesaurus of Euphemisms*. New York: McGraw-Hill, 1985.

Oxford English Dictionary. Oxford: Oxford University Press, 1971.

Partridge, Eric. *A Dictionary of Catch Phrases*. New York: Stein and Day, 1986.

———. *In His Own Words*. New York: Macmillan, 1980.

———. *Origins: A Short Etymological Dictionary of Modern English*. New York: Greenwich House, 1983.

———. *Smaller Slang Dictionary*. New York: Dorset, 1961.

Pei, Mario. *Double-Speak in America*. New York: Hawthorn, 1973.

Quinn, Jim. *American Tongue in Cheek*. New York: Penguin, 1982.

Radford, Edwin. *Unusual Words and How They Came About*. New York: Philosophical Library, 1946.

Rawson, Hugh. *A Dictionary of Euphemisms & Other Doubletalk*. New York: Crown, 1980.

Rogers, James. *The Dictionary of Cliches*. New York: Facts On File, 1985.

Roget, Peter Mark. *Roget's Thesaurus*. New York: Galahad, 1974.

Safire, William. *I Stand Corrected*. New York: Times Books, 1984.

———. *On Language*. New York: Times Books, 1980.

———. *What's the Good Word?* New York: Times Books, 1982.

The Second Barnhart Dictionary of New English. New York: Barnhart, 1980.

Sherk, William. *500 Years of New Words*. Toronto: Doubleday, 1983.

Shipley, Joseph. *Dictionary of Word Origins*. New York: Philosophical Library, 1945.

Tuleja, Tad. *Namesakes*. New York: McGraw-Hill, 1987.

Urdang, Laurence. *Names & Nicknames of Places & Things*. New York: Meridian, 1988.

Webster's New World Dictionary. New York: Simon & Schuster, 1986.

Why Do We Say It? Secaucus, N.J.: Castle, 1985.

Word Mysteries and Histories (by the editors of the *American Heritage Dictionary*). Boston: Houghton Mifflin, 1986.

Help!!!

HELP:
to give assistance,
to provide aid or support,
to serve, to rescue,
to lend a hand,
to remedy or
ameliorate!

Dear Dave,
Help! Why
do we say
"Help!"??

Signed,
In Need of
Assistance

So, what's the good word? Or phrase? What everyday expression have you been using without knowing why you say it? We are collecting more words and phrases for *Imponderables* and a possible sequel to this book.

If you are the first person to send in a question about a word or phrase we answer in any of our books, we'll send you a free copy of the book, complete with an obsequious acknowledgment of your contribution.

We'd also be interested in your comments about this book. If you would like a reply, a self-addressed, stamped envelope will assure one.

So send those words, phrases, and comments, along with your name, address, and telephone number, to:

WORD IMPONDERABLES
BOX 24815
LOS ANGELES, CA 90024

Index